# THE BIOFUEL BREAKTHROUGH

Industrial Nations Shift to Sustainable Energy, Biofuels and Climate Change Solutions (global warming books)

**Carina N Beger**

**Published by CB Storyworks.**

First Edition

ISBN 978-1-7644668-8-2 (eBook)

ISBN 978-1-7644668-7-5 (Paperback)

# About the Author

## Carina N Beger

Carina N Beger is a passionate advocate for sustainable energy and a respected voice in the biofuel industry. She brings a unique perspective shaped by two decades of experience in Contracts and Procurement in the Mining, Oil and Gas sectors. Her wealth of knowledge instills confidence in her readers.

Her book, "The Biofuel Breakthrough: Industrial Nations Shift to Sustainable Energy, Biofuels and Climate Change Solutions (global warming books)," is a comprehensive guide that explores cost-effective alternatives to fossil fuels. It focuses on the potential of biofuel solutions in addressing climate change and promoting sustainability.

Beyond merely presenting facts, Carina's work involves inspiring change. She meticulously researches every aspect of her subject matter, from clear regulatory guidelines and incentives to technological advancements to ensure compatibility.Furthermore, she is dedicated to giving readers scalable, environmentally sound infrastructure solutions, offering hope for a greener future.Carina sheds light on sustainable and ethical sourcing practices. She believes that understanding these practices is key to promoting their worldwide adoption.Her dedication to this cause shines through on every page she writes.Carina's work is a testament to her commitment to creating a more sustainable future on a global scale. She hopes her biofuel expertise will significantly influence the fight against climate change.Embark on your progress toward understanding the potential of biofuels with Carina N Beger now. Above all, Carina's vision extends beyond simply disseminating information. It involves fostering understanding and unity in our shared pursuit of emission reduction..

# ABSTRACT

*"The Biofuel Breakthrough:* Industrial Nations' Shift to Sustainable Energy, Biofuels and Climate Change Solutions (global warming books)" analyzes the global shift to sustainable biofuel technologies. It delves into the impact of technology, policy and markets on biofuel adoption. Readers will learn about implementation strategies, investment opportunities and regulatory frameworks for successful integration. Moreover, the book also provides practical advice for energy professionals to navigate the sustainable energy sector effectively.

# Contents

# Introduction

As the world struggles to address climate change, transitioning to sustainable energy has become more critical than ever. Biofuels emerge as a compelling solution amidst the myriad of renewable options available. Notably, biofuels offer a unique bridge between the current infrastructure and a sustainable future. Hydrotreated Vegetable Oil (HVO) is to reduce over 99% of Scope 1 emissions compared to fossil diesel.

As the global energy landscape stands at a pivotal crossroads, industrialized countries are moving away from fossil fuels and toward cleaner energy sources. This transition, however, is far from simple. Balancing technical, economic and regulatory complexities with operational efficiency and competitiveness is key.

However, sustainable energy transitions can reshape industries when implemented with care. Mining operations reduce their carbon footprint through carefully planned biofuel adoption, and watch as oil and gas companies diversify their portfolios to embrace renewable alternatives. A successful energy transition needs technology, policies, the right infrastructure and a well-functioning market.

This book is a comprehensive guide to the complexities of adopting and implementing biofuels. It covers new technology, important policies and ways to create sustainable supply chains. Actionable advice is available for all attendees, including industry professionals, policymakers, and sustainability managers.

Upcoming chapters discuss the main obstacles to biofuel use: high costs, poor infrastructure and supply problems. It finds cheaper ways to replace fossil fuels, explores compatible technologies, and offers clear rules and incentives. Most importantly, the book focuses on scalable infrastructure solutions with proven environmental impact.

Through real-world case studies and data-driven analysis, this book reveals how organizations successfully integrate biofuels into their operations while maintaining profitability and achieving sustainability goals. Specifically, focusing on mining, oil and gas, the book analyzes supportive policy frameworks, emerging markets, technological innovation and career opportunities in this expanding sector.

While this book highlights the promising role of biofuels in industrial decarbonization, it does not shy away from exploring limitations—economic, technical and environmental. The goal of emission reduction is not blind enthusiasm, but pragmatic optimism grounded in real-world data and experience.

Ultimately, the intention of this book is to serve as a guide to navigating the complexities of the biofuel revolution. It investigates how industrialized countries can make a successful transition to sustainable energy, making informed decisions and addressing the urgent need to address climate change. While the path ahead is challenging but exciting, with numerous opportunities for those ready to build a sustainable future.

Chapter 1

# FOUNDATIONS OF THE BIOFUEL REVOLUTION

## *Technology Landscape and Market Dynamics*

We are at a turning point for global energy as we transition from fossil fuels to renewable fuels. Biofuels bridge the gap between fossil fuels and a sustainable and environmentally friendly future. Cutting carbon emissions is crucial for heavy industry, but it must be done without impacting efficiency. Fortunately, biofuel technology has evolved, offering practical solutions for industrial decarbonization. These solutions can be used with existing infrastructure and pave the way for advanced applications.

The mining industry's work to lower carbon emissions from its heavy mobile equipment serve as a clear example of the path to cleaner energy. Initially, these efforts were met with skepticism, particularly from maintenance teams concerned about engine performance and reliability. However, transitioning to Hydrotreated Vegetable Oil (HVO) shows the potential of biofuels in industrial applications. Comprehensive monitoring systems can track fuel use, engine performance and emissions, and the results are striking. Interestingly, operational efficiency is unchanged despite substantial emission reductions, and notably, no major equipment changes were necessary.

Adopting biofuels depends more on willingness to change than on technological advancement. In fact, comprehensive documentation of performance metrics is important. It helps to overcome resistance and encourages wider adoption in the fleet. This experience exemplifies how small-scale trials can gradually catalyze large-scale transformations in industrial operations.

**Key Takeaways**

- Biofuels have evolved from simple crop-based systems to highly engineered, infrastructure-compatible fuels.

- Their adoption is already underway in sectors like mining and heavy transport, where drop-in fuels show measurable emission reductions with no major hardware changes.

- The successful integration of biofuels depends as much on stake-holder readiness and policy incentives as it does on technological maturity.

The above outlines the foundational interplay between biofuel technology and market readiness, setting the stage for deeper exploration of implementation, supply chain dynamics and investment strategy in the chapters ahead.

# Evolution of Biofuel Technologies: From First Generation to Advanced Solutions

Over the years, biofuel technology has undergone a remarkable transformation, evolving from rudimentary biomass combustion to sophisticated molecular and synthetic engineering. This journey is often categorized into generational phases, each representing a leap in ecosystem stewardship, efficiency and scalability. Understanding this evolution is crucial to appreciating the present-day capabilities of the biofuel industry and identifying the trajectories of future innovation.

### From Food Crops to Engineered Organisms: Generational Progress

The first generation of biofuels marked the beginning of large-scale production of alternative energy. Derived primarily from food crops such as corn, sugarcane and vegetable oils, these biofuels leveraged well-known agricultural processes and infrastructure. While they demonstrated feasibility at scale, their reliance on arable land and food resources raised concerns about food security and environmental sustainability.

In response to these limitations, second-generation biofuels emerged. These technologies use lignocellulosic biomass, including agricultural residues, forestry waste, and dedicated non-food energy crops, which are processed through enzymatic hydrolysis or thermochemical conversion. This generation significantly improved environmental performance by avoiding competition with food production and reducing waste.

The third generation advanced the field further by introducing microalgae and other high-yield organisms. Algal biofuels are particularly notable for their superior productivity, ability to grow on non-arable land and potential for carbon capture. These features position them as a sustainable solution for sectors with high energy demands, although economic scalability remains a challenge.

Fourth-generation biofuels represent the cutting edge of the industry, combining synthetic biology, advanced catalysis and genetic engineering. These systems aim to directly convert solar energy and carbon dioxide into usable fuels using engineered organisms and novel biochemical pathways. While largely in development, they hold the promise of a carbon-negative fuel cycle with minimal use of land and water.

**Feedstock Overview and Selection Criteria**

Feedstock selection is foundational to biofuel production. Each generation relies on distinct feedstock types:

- **First Generation:** Food crops $\rightarrow$ Ethanol, Biodiesel

- **Second Generation:** Agricultural residues, wood waste $\rightarrow$ Cellulosic Ethanol, Pyrolysis Oil

- **Third Generation:** Algae, cyanobacteria $\rightarrow$ High-yield lipids, Photosynthetic Fuels

- **Fourth Generation:** Engineered microbes + solar $CO_2$ capture $\rightarrow$ Synthetic Biofuels, carbon-negative tech

**Generations of Biofuel Technologies**

| First Gen | Second Gen | Third Gen | Advanced |
|---|---|---|---|
| Food Crops | Non-Food Biomass | Algae | Synthetic Biology |

The criteria for selecting appropriate feedstocks include:

- Availability

- Yield

- Environmental impact

- Land and water usage

- Processing efficiency and

- Compatibility with existing infrastructure

These factors influence not just fuel output, but also cost, supply chain resilience, and lifecycle emissions. As biofuel technologies advance, feedstock flexibility and sustainable sourcing have become increasingly critical for commercial viability.

## Technological Innovations and Process Enhancements

In parallel with generational improvements, significant innovations in processing technologies have emerged. Sophisticated catalytic systems, advanced fermentation techniques and real-time process controls have optimized conversion efficiencies and fuel quality. Artificial intelligence and machine learning further enhance operational precision, enabling adaptive systems that respond dynamically to variations in feedstock. Chapter 9 describes processing technologies in more detail.

A major leap in compatibility has come from "drop-in" biofuels—molecularly identical to fossil fuels, allowing for immediate integration into existing engines and infrastructure. These are particularly valuable in aviation, maritime transport and heavy industry, where electrification is not yet viable.

Modern production facilities now incorporate closed-loop systems, which significantly improve environmental metrics such as water use, waste management, and energy efficiency. These integrated approaches have reshaped the sustainability profile of biofuels, addressing long-standing criticisms regarding their ecological impact.

## Looking Forward

Emerging technologies promise to further elevate the biofuel industry. Advances in synthetic biology, modular processing and carbon capture integration are expected to yield more cost-effective, scalable and environmentally friendly solutions.

As energy systems transition globally, biofuels are poised to play an increasingly strategic role—not only as substitutes for fossil fuels, but as cornerstones of a sustainable and circular energy economy.

# Market Forces and Economic Drivers Shaping the Biofuel Industry

A complex interplay of economic incentives, regulatory frameworks, technological advancements and supply chain realities shapes the dynamic growth of the biofuel sector. A nuanced understanding of these forces is essential for stakeholders navigating this rapidly evolving landscape.

## Economic Fundamentals and Cost Dynamics

Biofuel competitiveness is closely tied to fossil fuel price volatility, carbon pricing mechanisms and renewable energy incentives. As carbon regulations tighten across jurisdictions, low-carbon fuels gain economic favor, especially in hard-to-decarbonize sectors such as aviation, shipping and heavy industry. Incentives, such as tax credits, grants and mandates, help bridge cost gaps and catalyze adoption.

At the core of production economics lies feedstock cost and availability. As outlined in chapter 3, feedstock properties influence not only energy output and environmental impact but also cost structure and supply chain complexity. Seasonal variability, land use pressures, and global commodity price fluctuations create uncertainty in production planning. Companies are increasingly seeking to mitigate these risks through diversified sourcing and investing in resilient, regional feedstock supply chains.

## Demand Drivers and Sectoral Momentum

Rising corporate emission reduction commitments have created stable demand signals for biofuels, particularly in aviation and logistics. Sustainable Aviation Fuel (SAF), for instance, has seen a surge in development driven by international emissions reduction targets and industry pledges. Algal-based and lignocellulosic fuels are among the frontrunners, offering attractive carbon intensity metrics and scalability potential.

In the automotive sector, regulatory mandates for renewable blending and emissions thresholds continue to shape demand, although electrification poses long-term questions about market share. However, for sectors where electrification is limited, biofuels remain one of the few scalable options for decarbonization.

## Technology-Driven Cost Reductions

Technological progress has substantially lowered production costs across multiple biofuel pathways. Innovations in enzymatic hydrolysis, catalyst efficiency and bioreactor design have improved yields and reduced operational complexity. Digitalization and automation enable precise control over inputs and outputs, minimizing waste and maximizing throughput.

These advancements are key to enabling parity with fossil fuels, especially as economies of scale are realized in commercial operations. However, cost competitiveness still often depends on policy mechanisms, particularly in early-stage markets.

## Policy and Regulatory Ecosystems

Policy remains a cornerstone of biofuel market development. Instruments such as the Renewable Fuel Standard (RFS), the Low Carbon Fuel Standard (LCFS), and international blending mandates provide essential economic signals to guide investment and development. Importantly, these policies also incentivize infrastructure readiness, including fuel terminals, blending facilities and dedicated distribution networks.

The design and stability of policy frameworks significantly influence investor confidence. Clarity around long-term goals, compliance requirements and support mechanisms reduces perceived risk and enhances capital inflow into commercial-scale projects.

## Infrastructure and Investment Trends

Infrastructure remains both a hurdle and an opportunity. Establishing end-to-end logistics—from feedstock collection to final delivery—requires coordinated investment. Yet, early movers in infrastructure development often secure competitive advantages as regional markets mature.

Recent investment trends indicate a shift from predominantly government-backed pilot programs to increased private sector engagement in large-scale operations. Venture capital, strategic partnerships, and public-private consortia are playing a greater role in funding innovative technologies and scaling production.

## Regional Markets and Global Variability

Biofuel adoption varies significantly by region, depending on local policy, resource availability, infrastructure and energy needs. For example, North America leads in corn- and soy-based ethanol, while Europe emphasizes waste-derived biodiesel and advanced biofuels. Emerging markets in Asia and Latin America offer significant potential, particularly where agricultural residues are abundant.

Understanding these regional nuances is essential for companies planning market entry or scaling operations. Feedstock availability, regulatory alignment and logistical feasibility must all be factored into strategy.

## Outlook: A Shifting Competitive Landscape

The biofuel industry's future will be shaped by consolidation, innovation and integration into broader energy systems. As carbon pricing intensifies and sustainability metrics become central to business models, biofuels will play an increasingly important role in the global decarbonization process.

Feedstock sourcing strategies will evolve toward circular economy principles, focusing on waste valorization, regional resilience and environmental stewardship. Companies that can balance technological innovation with operational agility and strategic partnerships will define the next phase of growth in this dynamic and competitive sector.

# The Diverse World of Biofuel Feedstocks

A comprehensive understanding of the variety of potential feedstocks is key to a successful biofuel strategy. In particular, careful evaluation is crucial when selecting raw materials for biofuel systems, since each offers distinct benefits and difficulties depending on the specific application. Moreover, the selection of suitable feedstocks significantly influences production costs, environmental impact and overall system sustainability.

Traditionally, agricultural feedstocks, including corn, sugarcane, and oilseed crops, have historically dominated first-generation biofuel production. While these materials benefit from well-established supply chains and conversion technologies, their use raises serious concerns regarding food security and land use. As a result, innovation has shifted toward alternative feedstock sources to help ease the conflict between fuel and food production.

One such innovation involving lignocellulosic biomass represents a significant advancement in feedstock utilization. This category includes agricultural residues, forestry waste and dedicated energy crops, all of which offer substantial advantages in terms of sustainability and resource efficiency. For instance, corn stover, wheat straw, and wood waste can be converted into biofuels while also providing additional value streams from existing agricultural and forestry operations.

**Primary Lignocellulosic Feedstocks:**

- Agricultural residues (corn stover, wheat straw)

- Forestry residues (logging waste, mill residues)

- Dedicated energy crops (switchgrass, miscanthus)

- Municipal solid waste

The emergence of advanced feedstock options has significantly expanded the possibilities for producing sustainable biofuels. Among these options, algae biomass stands out as a promising resource, offering high yields per acre and the potential for cultivation on land unsuitable for traditional agriculture. Moreover, these microscopic organisms can produce oils suitable for conversion into various types of biofuel while also capturing atmospheric carbon dioxide, contributing to efforts to reduce emissions.

In addition to biological sources, industrial waste streams are increasingly recognized as valuable feedstock sources, further supporting circular economy principles and, reducing waste disposal challenges at the same time. For example, used cooking oils, animal fats, and industrial byproducts can be effectively converted into high-quality biofuels, offering both environmental and economic benefits. Notably, this approach has proven particularly successful in regions with well-developed collection and processing infrastructure, demonstrating how existing waste management systems can be leveraged for sustainable energy solutions.

Furthermore, biofuel production strategies depend heavily on the geographical distribution of feedstock resources. Factors such as regional differences in climate, land availability and agricultural practices play a crucial role in shaping feedstock selection and supply chain development. Therefore, understanding these spatial dynamics is essential for developing robust and resilient biofuel production systems that can adapt to varying regional conditions.

**Key Feedstock Selection Criteria:**

- Local availability and supply stability

- Production and collection costs

- Environmental impact and sustainability metrics

- Processing requirements and conversion efficiency

As innovation progresses, emerging technologies are expanding the range of viable feedstock options. Specifically, advanced preprocessing techniques and novel conversion methods make previously challenging materials accessible for biofuel production. These advancements are particularly crucial for utilizing heterogeneous waste streams and improving the efficiency of existing feedstock processing.

Beyond technological feasibility, the sustainability implications of feedstock choice extend beyond immediate environmental impacts. In particular, considerations must include soil health, water usage, biodiversity effects and long-term ecosystem stability. To address these complexities, effective feedstock strategies typically use diverse sources to balance various factors and ensure a reliable supply.

At the same time, cost considerations remain paramount in feedstock selection. Key factors, such as transportation distance, storage requirements, and preprocessing needs, significantly impact the overall economics of biofuel production.

Moving forward, digital technologies and advanced analytics will increasingly improve feedstock management and supply chain optimization. For instance, real-time monitoring and predictive analytics keep supplies steady while simultaneously cutting waste and storage expenses. As a result, these tools are becoming increasingly important as biofuel operations scale up to meet growing demand.

At the same time, the range of biofuel feedstocks is constantly expanding, thanks to the discovery of new sources and improvements in processing methods. To succeed in this evolving landscape, careful evaluation of available options, consideration of local conditions and strategic planning are essential to ensure sustainable and economically viable operations. Therefore, understanding these dynamics is crucial for stakeholders who seek to develop or expand their biofuel production capabilities.

In parallel with these developments, the technological landscape of biofuel production continues to evolve rapidly, driven by innovations in processing methods, catalyst development and system integration. These breakthroughs are not only revolutionizing sustainable fuel production but also overcoming past obstacles to efficiency and scalability.

The development of advanced enzymatic systems for biomass conversion has led to a significant breakthrough in processing technology. Specifically, these biological catalysts have dramatically improved the efficiency of converting lignocellulosic materials into fermentable sugars, thereby addressing a longstanding bottleneck in second-generation biofuel production. Moreover, optimizing these enzyme cocktails has reduced processing times and lowered production costs, ultimately making cellulosic biofuels increasingly competitive with conventional alternatives.

In addition to enzymatic advancements, thermochemical conversion technologies have also made significant progress, particularly in gasification and pyrolysis processes. Notably, modern systems achieve higher conversion efficiencies while simultaneously producing fewer unwanted byproducts, marking a substantial improvement over earlier generations of technology. Furthermore, integrating advanced control systems and real-time monitoring has enabled more precise process control, resulting in consistent product quality and lower operational costs.

## Key Technological Advances:

- Enhanced enzymatic conversion systems

- Improved thermochemical processing

- Advanced catalyst formulations

- Integrated process control systems

Catalyst technology, in particular, has played a pivotal role in improving biofuel production efficiency. Specifically, new catalyst formulations enhance reaction selectivity, lower energy requirements and enhance operational lifetimes due to greater stability, thereby reducing maintenance needs. These advancements have been especially crucial in hydro processing applications, where catalyst performance directly impacts product quality and process economics.

Beyond catalyst improvements, another significant advancement is integrating artificial intelligence and machine learning into process control systems. By enabling real-time optimization of operating parameters, predictive maintenance scheduling and improved quality control, these technologies enhance overall efficiency. Moreover, handling large operational datasets provides valuable insights that drive ongoing improvements in efficiency and product quality.

At the same time, process intensification strategies are focusing on combining multiple unit operations into smaller, more efficient systems. By integrating these designs, industries can reduce capital costs, improve energy efficiency and minimize waste generation. Additionally, modular systems enhance scalability, allowing for flexible deployment across various production units.

**Process Integration Benefits:**

- Reduced capital and operating costs

- Improved energy efficiency

- Enhanced product quality control

- Greater operational flexibility

In recent years, innovative hybrid systems that blend biological and thermochemical conversion methods have emerged as a novel approach to maximizing resource use. By combining these techniques, such systems can handle a broader range of feedstocks while simultaneously producing multiple value-added products, thereby improving overall economic viability.

In addition to hybrid conversion methods, waste heat recovery and energy integration systems have become increasingly sophisticated, leading to significant improvements in the overall energy efficiency of biofuel production facilities. For instance, optimizing heat exchanger networks and thermal integration not only lowers energy consumption but also reduces environmental impacts, making production more sustainable.

Meanwhile, membrane separation technologies have advanced significantly, offering more efficient methods for product purification and concentration. Compared to two additional techniques, these systems use less energy while also producing purer products at higher recovery rates. As a result, they contribute to both cost reduction and improved biofuel quality.

Looking ahead, electrochemical conversion and photobiological systems are promising technologies for future development. Although still at various stages of research, these technologies have the potential to offer new pathways for sustainable fuel production while simultaneously reducing environmental impact.

As these innovations progress, integrating these technological advances has created opportunities for significant improvements in biofuel production efficiency and sustainability. Nevertheless, successful implementation requires careful evaluation of specific operational requirements and local conditions. Effective biofuel production requires a thorough understanding of technology's capabilities and limitations. In this context, effective biofuel production depends on a thorough understanding of technology's capabilities and limitations.

# Stakeholder Dynamics in Biofuel Transitions

Successfully advancing biofuel adoption requires more than technological readiness—it hinges on the coordinated involvement of a diverse range of stakeholders, each with distinct motivations, constraints and roles. Understanding these dynamics is essential to developing targeted engagement strategies that build trust, ensure operational success, and foster long-term support.

## Contested Futures: Diverse Perspectives on Biofuel Development

While biofuels offer compelling solutions for decarbonization, their adoption is not without debate. Stakeholders across the academic, environmental and social sectors have raised legitimate concerns about large-scale biofuel deployment. Including these perspectives ensures a more nuanced, credible and balanced understanding of what is at stake.

## Key Stakeholder Groups and Their Motivations

| Stakeholder Group | Motivation & Priorities | Common Concerns or Barriers |
|---|---|---|
| Farmers & Feedstock Producers | Stable income, inclusion in new value chains, resilient rural development | Market volatility, land-use conflict, upfront equipment costs |
| Investors & Project Developers | Return on investment, policy certainty, ESG alignment | Regulatory risk, high capital intensity, unclear credit timelines |
| Local Communities | Employment opportunities, health and environmental co-benefits | Land use impact, transparency, lack of participation |
| Industrial End-Users | Fuel reliability, drop-in compatibility, emissions compliance | Retrofitting costs, operational downtime, fuel quality assurance |
| Policymakers & Regulators | Climate targets, energy security, economic development | Budget constraints, public accountability, enforcement capacity |
| Environmental NGOs | GHG reduction, biodiversity protection, ethical land sourcing | Food vs fuel debate, monitoring credibility, traceability gaps |
| Fuel Blenders & Distributors | Streamlined logistics, infrastructure efficiency, regulatory compliance | Storage compatibility, blend variability, liability concerns |

## Tailored Engagement Strategies

- **Farmers:** Offer guaranteed feedstock offtake agreements, access to extension services and training in sustainable practices.

- **Investors:** Provide transparent risk assessments, hybrid financial models (e.g., green bonds + offtakes) and strong policy alignment.

- **Communities:** Establish local consultation bodies, community development funds and employment pipelines.

- **End-Users:** Showcase case examples with proven performance metrics, phased integration and emissions data.

- **Policymakers:** Deliver scenario modeling, international best-practice comparisons and lifecycle impact studies.

- **NGOs:** Incorporate third-party certification schemes, land-use safeguards and grievance redress mechanisms.

- **Distributors:** Coordinate with regulators on infrastructure standards and implement digital quality control tracking.

> Cities are the front lines of climate action. Biofuels offer a way to de carbonize fleets without waiting for perfect infrastructure." – Thomas Leong, Urban Sustainability Advisor

## Critical Viewpoints to Consider

| Perspective | Core Argument | Example Concerns |
|---|---|---|
| Environmental NGOs | Caution against indirect land-use change (ILUC) and biodiversity loss | Forest clearing for energy crops, monoculture expansion |
| Food Security Advocates | "Fuel vs. Food" concerns when arable land or crops are diverted from food to energy | Spike in food prices, especially in lower-income countries |
| Climate Skeptics & Analysis | Challenge life-cycle emissions accounting and overreliance on optimistic modeling | Do current methods truly outperform fossil fuels over time? |
| Local Communities | Demand inclusion, transparency and equitable benefit-sharing | Land access disputes, broken promises on job creation |
| Supply Chain Experts | Warn about the scalability, quality control and fraud risk in biofuel certification | Market flooding with uncertified or low-grade product |

## Strategies for Inclusion and Responsiveness

- **Participatory Impact Assessments:** Go beyond regulatory compliance by integrating local voices early in feasibility studies.

- **Transparent Life Cycle Analysis:** Acknowledge assumptions in emissions modeling and engage independent reviewers.

- **Social Safeguards in Project Design:** Prioritize equitable land-use policies, community development funds and local hiring mandates.

- **Platform for Dissenting Views:** Encourage constructive criticism through open data, policy hearings and collaborative forums.

> *"Our forests aren't just carbon sinks—they're homes, water sources, and ancestral land. We weren't consulted before this biofuel project started."*
>
> *— Lilian Mapendo, Community Leader, Eastern DR Congo*

## Meet the Experts: Voices Driving the Biofuel Transition

To enhance narrative depth and credibility throughout the book, we introduce four recurring expert personas. Each brings a unique lens to biofuel adoption, bridging the gap between technology, policy and practice.

| Expert Name | Role & Background | Chapter Presence |
|---|---|---|
| Dr. Nia Rahman | Chemical engineer with 15+ years in renewable fuels; leads pilot-scale bio-refinery program | Chapters 2 & 9 (Technical analysis, scale-up risk) |
| Luis Ortega | Procurement director at a multinational mining firm, championing green supply chains | Chapters 3 & 6 (Supply Chain, Stakeholder Engagement) |
| Anya Becker | Independent ESG investor and former energy market analyst | Chapters 4 & 10 (Investment modeling, strategic planning) |
| Thomas Leong | Urban sustainability advisor and former city council member | Chapters 1 & 5 (Public outreach, regulatory alignment) |
| Minister Karima Desai | Deputy Minister for Renewable Energy and Climate Affairs | Chapters 5 & 6 (Policy, Stakeholder engagement) |
| Lilian Mapendo | Community Leader, Eastern DR Congo | Chapter 1, (Stakeholder dynamics) |

As we conclude this foundational chapter, the critical role of biofuels in potentially transforming our energy landscape becomes increasingly apparent. Indeed, technological advancements, market shifts, and varied feedstocks have collectively shown the biofuel industry's vast potential while also highlighting the considerable challenges that need to be addressed.

Throughout this discussion, we have seen impressive progress in biofuel technology, evolving from rudimentary first-generation systems to cutting-edge advanced biofuels. With each technological generation, improvements in efficiency, sustainability, and commercial viability have been achieved, while also addressing the limitations of earlier systems. The diverse world of feedstock options presents both opportunities and challenges for the biofuel industry. We've also acknowledged that there are stakeholders with different interests that need to be taken into account.

## Key Takeaways

- Biofuels are already proving viable in heavy industry, particularly mining, through real-world success with HVO.

- Technological compatibility (e.g., drop-in fuels) makes them immediately adoptable without a significant overhaul of infrastructure.

- Key success factors: data-driven pilot programs, stakeholder dynamics and addressing human factors.

In the following chapters, we will expand on detailing strategies for technology selection, supply chain development, effective implementation and policies. Indeed, the journey toward sustainable energy solutions through biofuels is complex yet entirely achievable, provided it is guided by careful planning, stakeholder engagement and continuous innovation.

Chapter 2

# Strategic Assessment

## *Evaluating Biofuel Technologies for Industrial Applications*

This chapter is designed primarily for industrial energy profession-als—engineers, operations managers and project developers—tasked with evaluating and deploying biofuel technologies. However, several key insights may also assist policymakers involved in infrastructure planning and industrial decarbonization incentives. Sections explicitly related to policy and regulatory impact are flagged for dual relevance.

The successful implementation of biofuel technologies in industrial settings demands more than technical feasibility. Rather, it requires a strategic approach considering the entire production, distribution and application ecosystem. Assessing biofuel technologies requires a balanced approach that takes into account technical, economic and environmental factors. Specifically, industry leaders must carefully evaluate feedstock availability, processing requirements and compatibility with existing systems while also keeping long-term emissions reduction goals in focus.

## Biofuel Technology Evaluation Matrix

| Criteria | Corn Ethanol | Cellulosic Ethanol | HEFA Diesel | Synthetic Gas Fermentation |
|----------|-------------|--------------------|-------------|----------------------------|
| Feedstock Availability | High | Moderate | High | Variable |
| Infrastructure Compatibility | High | Medium | High | Low |
| GHG Reduction Potential | Moderate | High | High | Very High |
| CapEx Intensity | Low | High | Medium | High |
| Scale-Up Maturity | High | Medium | High | Low |

This framework enables comparison of technologies not solely on lab performance but on how well they integrate with existing systems. A high GHG reduction potential may be negated by a poor infrastructure fit or prohibitively high costs.

Furthermore, the integration of renewable alternatives into current operations poses a significant challenge to a strategic assessment. To address this, evaluating regional feedstock availability, processing capabilities, storage and equipment compatibility necessitates a comprehensive assessment matrix. The most technologically advanced option is not always the optimal choice. In many cases, a well-established, simpler biofuel solution can provide greater reliability and cost-effectiveness for industrial operations.

The assessment process must consider multiple dimensions of implementation, ranging from technical feasibility to economic viability.

**Key evaluation criteria may include:**

- Current capability

- Direct emissions reductions

- Broader lifecycle effects

- Local infrastructure

- Supply chain

- Operational flexibility

- Business continuity

- Future scalability

To achieve a well-rounded analysis, a robust evaluation framework should incorporate both quantitative metrics and qualitative factors such as:

- Energy density

- Carbon intensity

- Conversion efficiency

*"Feedstock inconsistency is still the biggest optimization hurdle,"* *warns Dr. Nia Rahman.Biorefinery Pilot Program Lead*

By adopting this balanced approach, organizations avoid the pitfall of focusing solely on technical specifications while inadvertently overlooking practical implementation challenges.

### Biofuel Technology Evaluation Matrix

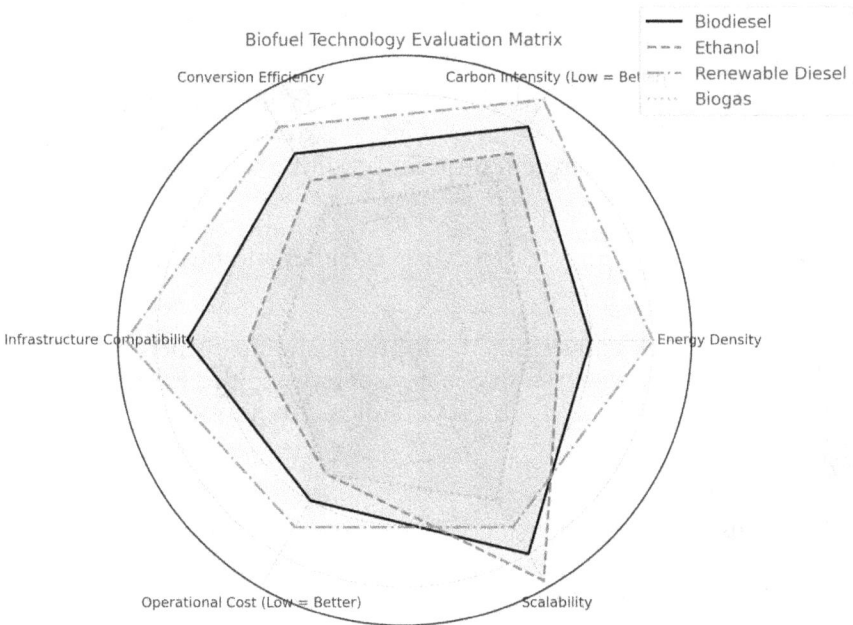

Biofuel Technology Evaluation Matrix

Legend:
— Biodiesel
--- Ethanol
—··— Renewable Diesel
····· Biogas

Axes: Conversion Efficiency, Carbon Intensity (Low = Better), Energy Density, Scalability, Operational Cost (Low = Better), Infrastructure Compatibility

Through systematic evaluation and strategic planning, organizations can identify solutions that not only meet current emission reduction goals but also provide a foundation for long-term operational success. Ultimately, the key is maintaining a balanced perspective that accounts for both immediate practical requirements and future strategic objectives.

## Technical Feasibility and Infrastructure Requirements for Industrial Biofuel Integration

When evaluating the integration of industrial biofuels, seven fundamental pillars determine the technical feasibility:

- Infrastructure compatibility

- Storage capacity and infrastructure

- Distribution networks

- Maintenance and implementation disruption

- Safety systems and spill control upgrades

- Fuel transport and blending protocol

- Space

**Pro Tip (Energy Professional Focus):** For facilities with legacy fossil infrastructure, HEFA diesel may provide a faster return due to drop-in compatibility, while gas fermentation for cellulosic systems requires extensive retrofitting.

**Note to Policymakers:** Infrastructure subsidies and blending mandates can help de-risk these transitions.

To ensure a smooth transition, a comprehensive assessment of existing industrial systems is essential in order to determine the scope of modifications required for successful biofuel implementation.

**Key factors to be evaluated:**

- Material compatibility

- Equipment specifications

- Operational parameters

to ensure a potential seamless integration while maintaining operational efficiency.

Moreover, infrastructure requirements vary significantly depending on the type of biofuel being implemented.

**Modifications / Upgrades may be needed**

- Storage tanks

- Seals

- Fuel lines

- Filtration systems to handle contamination issues

- Temperature control systems to manage viscosity variations

When using biodiesel, due to its unique chemical characteristics.

Unlike conventional fossil fuels, many biofuels are hygroscopic and can absorb atmospheric moisture, which can potentially degrade their quality over time. Therefore, maintaining fuel quality necessitates storage facilities that are properly equipped for moisture and temperature control, as well as real-time monitoring systems. Furthermore, storage systems should be designed to effectively accommodate seasonal changes in feedstock supply and demand, ensuring both efficiency and reliability.

## Schematic Biofuel Storage System

### Schematic: Biofuel Storage System

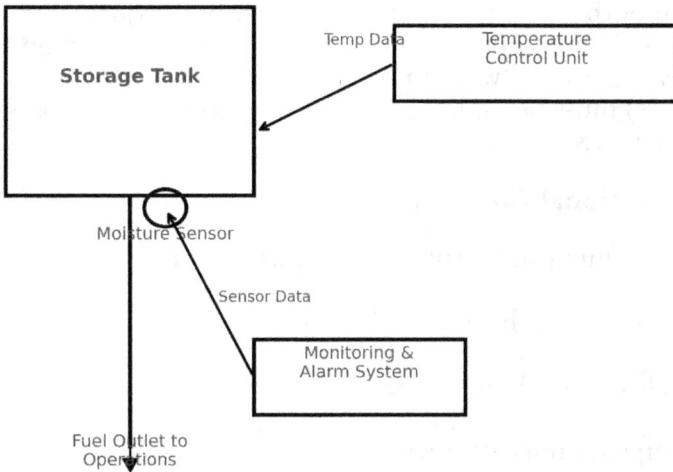

This diagram illustrates a standard storage setup for biofuels, highlighting key components necessary to maintain fuel quality and operational reliability. The system includes a corrosion-resistant storage tank, an integrated temperature control unit to manage viscosity and prevent degradation, a moisture sensor to monitor humidity intrusion and a central monitoring and alarm system for real-time oversight. A direct outlet connects the storage unit to operational systems, ensuring controlled distribution.

## Key Infrastructure Requirements:

- Fuel handling and transfer systems

- Temperature control and monitoring equipment

- Corrosion-resistant storage tanks and piping

- Enhanced filtration and purification systems

- Emergency containment and safety equipment

The infrastructure design must safeguard product integrity while also adhering to handling and transportation regulations. In addition to storage considerations, the successful integration of biofuels also depends heavily on the availability and reliability of distribution networks, which is covered in chapter 3.

To ensure smooth transitions, personnel must be adequately trained to handle new fuel types, operate modified equipment and respond to potential issues. Additionally, documentation and Standard Operating Procedures (SOPs) must be updated to accurately reflect new processes and safety requirements.

## Critical Operational Considerations:

- Staff training and certification requirements

- Maintenance schedule modifications

- Quality control procedures

- Emergency response protocols

- Performance monitoring systems

As biofuel technology continues to advance, infrastructure design should incorporate flexibility for future upgrades and modifications. A proactive strategy protects infrastructure investments while also ensuring adaptability to changing market conditions and regulatory landscapes.

## Economic Analysis: Cost-Benefit Assessment of Biofuel Technology Implementation

To make sound investment decisions, a full economic analysis of biofuel technology must weigh its direct and indirect costs against the potential benefits and long-term value.

**Cost Considerations:**

- Capital spending for retrofitting (e.g., biodiesel filtration systems, seal replacements)

- Operating costs biofuel vs fossil fuels

- Carbon credit and LCFS eligibility

- Projected revenue

- Resilience to feedstock market volatility

- Government incentives

- Equipment modifications

- Storage infrastructure

- Infrastructure modification and equipment upgrades

- Staff training and certification programs

- Supply chain development and logistics

- Quality control and monitoring systems

- Regulatory compliance and certification costs

Long-term operational benefits, such as reduced maintenance costs and potential carbon credit revenues, can significantly offset these upfront expenses.

By conducting a thorough cost-benefit analysis, organizations can make informed decisions about biofuel adoption while simultaneously managing financial risks effectively.

## Cost Breakdown of Biofuel Implementation

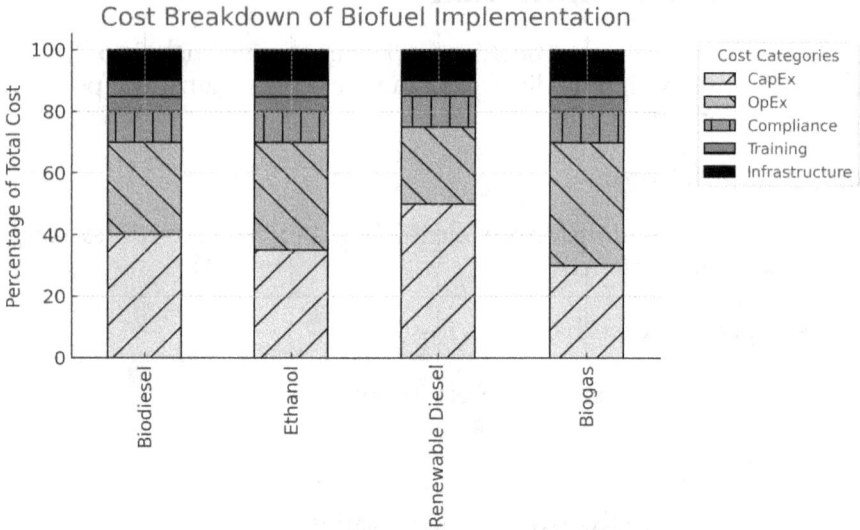

Cost Breakdown of Biofuel Implementation

Beyond direct operational savings, the benefits of biofuel implementation extend into broader financial and strategic areas. Specifically, businesses should factor in various incentives, including:

- Environmental credits

- Tax breaks

- Higher prices for eco-friendly goods

Moreover, demonstrating environmental stewardship can also enhance corporate reputation and stakeholder relationships. However, these intangible benefits may be more challenging to quantify in traditional financial terms.

At the same time, fluctuations in feedstock prices, processing efficiency, and maintenance costs must be factored into the operational cost analysis. Although some biofuels may have higher per-unit costs than conventional fuels, improved engine performance and reduced maintenance needs can help offset these expenses. Additionally, developing local supply chains can also provide long-term cost advantages while also supporting regional economic development.

**Key Benefit Categories:**

- Reduced carbon emissions and environmental compliance costs

- Potential tax incentives and government subsidies

- Enhanced energy security and supply chain resilience

- Improved corporate sustainability metrics

- Market differentiation opportunities

In any economic analysis, risk assessment plays a crucial role. To ensure financial stability, organizations must evaluate potential disruptions, such as feedstock supply, price volatility and regulatory changes that could impact the financial viability of biofuel projects. Implementing substantial risk mitigation, including diversifying suppliers and maintaining flexible production capabilities, helps safeguard investments and ensures long-term viability.

Additionally, the timing of implementation can significantly impact economic outcomes. For many organizations, a phased approach often proves beneficial, as it allows them to optimize their investment strategy while simultaneously building operational experience. Furthermore, this gradual implementation enables better management of capital expenditure while also providing opportunities to leverage lessons learned from initial rollouts.

## Phased Biofuel Implementation Timeline

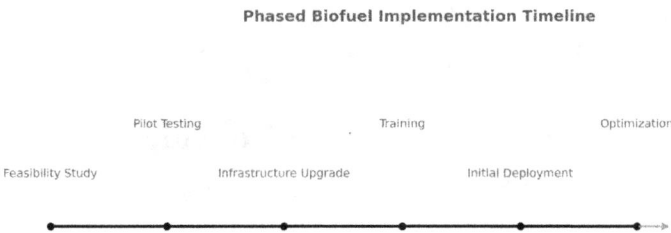

**Phased Biofuel Implementation Timeline**

| Pilot Testing | Training | Optimization |
| Feasibility Study | Infrastructure Upgrade | Initial Deployment |

Not all biofuel technologies scale evenly.

**Strategic implementation involves:**

- Matching technology maturity to project risk appetite

- Phased implementation aligned to capex cycles

- Staff training and ISO compliance timelines

Another critical aspect of economic assessment is the implementation of monitoring and verification systems. To accurately measure actual costs and benefits against projections, organizations must establish robust tracking mechanisms. Moreover, data analysis not only helps identify areas for optimization but also provides valuable insights to guide future growth and strategic decision-making.

The economic viability of biofuel implementation varies significantly across industries and regions. In particular, project feasibility depends on local factors, such as resource availability, infrastructure and regulations. Therefore, organizations must carefully evaluate these contextual factors when developing their economic assessment frameworks.

To enhance decision-making, economic analyses should incorporate long-term scenario planning to anticipate potential changes in technology, regulations and market conditions. By adopting this forward-thinking approach, organizations can make sound investment decisions in response to our changing energy landscape. Furthermore, considering multiple scenarios enables more robust planning while simultaneously highlighting potential risks and opportunities.

Moreover, successful economic assessment requires collaboration across multiple organizational functions, including operations, finance and sustainability teams. This integrated approach ensures that all relevant factors are considered while also promoting alignment between technical requirements and financial objectives. Additionally, regularly updated economic models help maintain relevance in changing markets and align with evolving company goals.

## Risk Management and Compliance Considerations in Industrial Biofuel Adoption

Implementing biofuel technologies in industrial settings introduces a complex matrix of risks and compliance requirements that must be carefully addressed to ensure successful adoption. However, the scope and complexity of risk management in a biofuel transition are often underestimated by organizations. To effectively navigate these challenges, a robust risk management system needs to cover operational, financial, regulatory and environmental factors, while also keeping up with changing compliance rules.

In addition to this, successful integration of biofuels necessitates meticulous handling of operational risks, such as quality control, storage, and equipment compatibility. Biofuels are known to be more susceptible to environmental factors compared to traditional fuels, thus demanding improved monitoring and control systems. Factors like temperature fluctuations, moisture content and microbial growth need to be closely monitored to uphold fuel quality and safeguard equipment from damage. To address these challenges, organizations should establish strong quality assurance protocols and conduct regular testing procedures.

**Key Operational Risks:**

- Fuel quality degradation

- Equipment compatibility issues

- Storage and handling complications

- Supply chain disruptions

- Performance variability

Various regulatory frameworks, including those related to environmental, safety, and transportation, govern the adoption of industrial biofuels. Organizations are confronted with a complex network of regulations at the local, national, and international levels, necessitating thorough documentation and reporting. Through regular audits and compliance assessments, organizations can ensure adherence to regulatory standards and proactively address potential areas of concern before they escalate into significant issues.

Supply chain risk management poses particular challenges in biofuel adoption. To address this, businesses need to establish rigorous supplier qualification procedures and maintain a variety of supply sources to ensure consistent availability. Quality control must be maintained across the entire supply chain, from the sourcing of raw materials to final product delivery.

Clear specifications and performance requirements are essential for the consistent evaluation of suppliers. Managing the financial risks of biofuel adoption requires close attention to market fluctuations, pricing strategies and possible regulatory shifts.

Organizations should develop hedging strategies to protect against price fluctuations and maintain adequate insurance coverage for potential operational disruptions. Additionally, long-term contracts and strategic partnerships can help stabilize costs and provide additional security in supply arrangements.

**Critical Compliance Areas:**

- Environmental permits and certifications

- Safety protocols and training requirements

- Transportation and handling regulations

- Emissions monitoring and reporting

- Quality control documentation

Effective stakeholder management is a crucial aspect of risk mitigation. To encourage biofuel adoption, organizations must communicate effectively with employees, regulators, local communities and other stakeholders, addressing their concerns and managing expectations. Maintaining clear communication channels and providing regular updates helps to maintain transparency and build trust among stakeholder groups.

It is also important to update emergency response planning to address the specific risks associated with biofuel systems. This includes developing comprehensive incident response procedures and ensuring that emergency teams are adequately trained to handle biofuel-related incidents. Conducting regular drills and simulations can help maintain readiness and identify potential gaps in emergency response capabilities.

As organizations implement new monitoring and control systems for biofuel operations, technology risk management becomes increasingly important. Improved cybersecurity is essential for safeguarding critical infrastructure and ensuring the reliability of monitoring and reporting systems. Regular updates and vulnerability assessments are needed to uphold security and comply with changing standards.

To effectively identify, prioritize, and mitigate risks, organizations can utilize a thorough risk register. This living document should be regularly reviewed and updated to reflect changing conditions and emerging risks. Performance metrics and key risk indicators offer early warning of potential issues and enable proactive risk management.

## Risk Assessment Heatmap for Industrial Biofuel Adoption

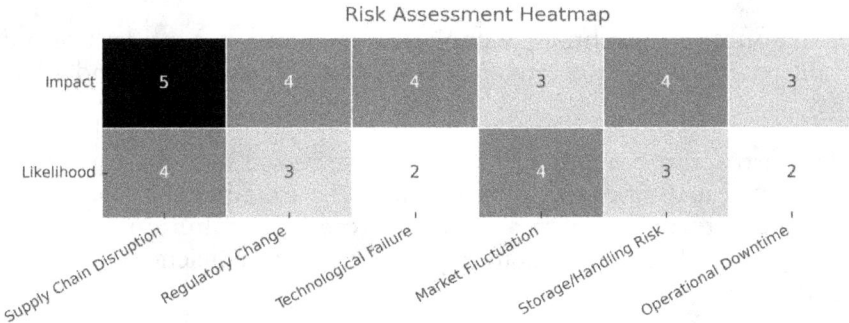

Risk Assessment Heatmap

| | Supply Chain Disruption | Regulatory Change | Technological Failure | Market Fluctuation | Storage/Handling Risk | Operational Downtime |
|---|---|---|---|---|---|---|
| Impact | 5 | 4 | 4 | 3 | 4 | 3 |
| Likelihood | 4 | 3 | 2 | 4 | 3 | 2 |

The matrix evaluates key operational and strategic risks based on their relative likelihood and impact. Darker shades represent higher criticality. This visual aids in identifying priority areas for mitigation planning during biofuel system integration.

Robust compliance monitoring and reporting systems are crucial for navigating complex regulations and providing stakeholders with timely, accurate data. Organizations benefit from implementing automated monitoring systems to reduce human error and ensure consistent data collection. Regular compliance audits help identify potential issues early and provide opportunities for continuous improvement.

As technology evolves and regulatory requirements change, these systems must adapt while maintaining effectiveness. Consistent review and updates on risk management strategies ensure alignment with company goals and regulations. The strategic assessment of biofuel technologies for industrial applications represents a critical juncture in the broader energy transition landscape.

The chapter outlines that successful implementation requires a systematic strategy that considers technical practicality, financial sustainability and environmental responsibility. Organizations can address the challenges of adopting biofuels by utilizing comprehensive assessment frameworks, as demonstrated by the case examples and analytical tools provided.

Assessing biofuel technology is a complex process that involves careful consideration of infrastructure, operations and regulations. Exploration of cost-benefit analyses and risk management strategies underscores the significance of thoroughly evaluating potential solutions. A matrix-based assessment method, which combines quantitative metrics with qualitative factors, offers a structured approach to evaluating biofuels within their specific industrial contexts.

To make successful technology choices, it is crucial to consider not only current technical requirements but also future sustainability and adaptability.

The infrastructure assessment framework discussed systematically evaluates existing capabilities and identifies necessary modifications for successful biofuel integration, with a focus on storage, distribution and monitoring systems. These systems are key to operational efficiency and require careful consideration.

It is important to note that analysis may reveal that the most advanced technological solution is not always the optimal choice for every industrial setting. Successful biofuel implementation is highly dependent on factors specific to each context, such as the availability of local feedstock, pre-existing infrastructure and workforce expertise. This reality underscores the importance of thorough pre-implementation assessment and strategic planning.

Technological progress, evolving regulations and shifting market forces will continue to influence the future of industrial biofuels. In order to adapt to new developments, organizations must maintain flexibility in their assessment frameworks and build robust evaluation processes.

By maintaining a balanced approach to technology evaluation that considers both immediate practical requirements and long-term strategic objectives, organizations can develop implementation strategies that meet current sustainability goals and provide a foundation for future growth and adaptation. This includes ensuring that biofuel investments provide long-term value and support sustainability.

---

■ *Reality Check: Scalability vs. Site-Specific Feasibility*

*While biofuel technologies may appear technically compatible on paper, real-world integration is often constrained by legacy infrastructure, regulatory bottlenecks and regional feedstock dynamics. For example, high-moisture feedstocks like sugarcane bagasse require specialized drying or preprocessing equipment not viable in arid or remote industrial zones. Additionally, not all facilities can easily retrofit systems for advanced bio-oil or ethanol blends without significant downtime or capital outlay.*

*Key Watchpoints:*

- *Seal compatibility and temperature control needs for biodiesel retrofits*
- *Varying feedstock economics by geography and season*
- *Local permitting delays and retrofit compliance timelines*

---

## Key Takeaways

- Implementation planning must begin with real-time baseline assessments across technical, operational, and financial domains.

- Success hinges on aligning technology choices with business models, particularly in sectors like mining or aviation, where infrastructure is already fixed.

- Strategic flexibility is essential; pre-approved contingency plans can mitigate common disruptions in biofuel adoption.

This chapter equips energy professionals and investors with an adaptive framework for navigating industrial transformation without compromising performance.

Chapter 3

# Sustainable Supply Chain Development

## *Infrastructure and Resource Management*

The backbone of any successful biofuel operation lies in its supply chain infrastructure, where every link must be carefully engineered to support sustainable practices while maintaining operational efficiency. Creating a resilient and environmentally conscious supply chain requires a delicate balance between optimizing resource utilization and minimizing environmental impact, while ensuring economic viability.

Developing sustainable supply chains has been challenging but rewarding, based on a long history in mining and energy. The transformation of traditional energy infrastructure into environmentally conscious systems requires careful consideration of multiple interconnected elements - from feedstock sourcing to final distribution networks.

Reliable and ethical feedstock sourcing is crucial for sustainable supply chains, which involves establishing strong relationships with suppliers, implementing quality control measures, and creating transparent tracking systems to verify the environmental impact of each supply chain component. Supply chain failure can lead to disruptions in supply and inconsistencies in quality.

Moving forward, this chapter focuses on sustainable biofuel supply chains, including infrastructure, resource management, logistics and digital integration. It also delves into the challenges of optimizing the supply chain, such as sourcing, transport, and storage, while emphasizing emission reduction and product quality.

## Ethical and Community Impacts of Biofuel Supply Chains

The pursuit of sustainable energy through biofuels brings with it a complex web of social and ethical considerations. While biofuels are often framed as a clean alternative to fossil fuels, their production and distribution can have unintended consequences for communities, ecosystems and food systems—particularly in low- and middle-income countries. This section explores these implications and identifies pathways for more equitable and responsible biofuel deployment.

## Land Use Change and Displacement

One of the most controversial aspects of biofuel expansion is its impact on land use. Converting land for energy crops can lead to deforestation, habitat destruction, and the displacement of indigenous and rural communities. A 2021 report by the FAO found that large-scale jatropha and sugarcane plantations in Madagascar and Mozambique displaced smallholder farmers and led to reduced food crop cultivation.

### Case Example: Soy Biodiesel Expansion in Brazil
*In the Brazilian Cerrado, soy plantations intended for biodiesel production have contributed to deforestation and encroachment on traditional territories. While the biofuel industry has created jobs, it has also driven land speculation and reduced access to land for subsistence farmers. Efforts by civil society groups have pushed for land tenure recognition and environmental safeguards, but enforcement remains inconsistent.*

## Food vs. Fuel Debate

When agricultural land is diverted for energy crops, it can reduce the availability of land for food production. This tension—often referred to as the "food vs. fuel" debate—becomes acute in regions where food security is already precarious. The 2008 global food price spike was partially attributed to increased demand for corn ethanol in the United States.

**Policy Insight:** Brazil and Thailand have developed biofuel zoning strategies to limit cultivation in areas vital for food production, setting an example of how to mitigate these tradeoffs through regulation.

## Water Use and Biodiversity Impacts

Biofuel crop production often demands intensive irrigation, which can strain local water systems and reduce water availability for domestic and agricultural use. Additionally, monoculture plantations reduce biodiversity, degrade soils, and increase pesticide reliance.

### Case Example: Palm Oil in Southeast Asia
*In Indonesia and Malaysia, oil palm plantations for biodiesel have replaced peatland forests, releasing stored carbon and threatening endemic species like the orangutan. Roundtable initiatives like RSPO have sought to certify sustainable palm oil, but adoption and compliance remain uneven.*

## Community Benefits and Energy Justice

While many communities bear the environmental burdens of biofuel production, fewer enjoy its benefits. Energy justice calls for a fair distribution of both the costs and benefits of energy transitions. Locally owned biofuel cooperatives—such as those in parts of Kenya and India—offer promising models.

### Case Example: Community-Scale Bioethanol in Kenya
*In Kitui County, Kenya, smallholder farmers produce bioethanol from cassava through a local cooperative. The initiative has provided off-grid fuel access, reduced deforestation and increased household income, illustrating how decentralized approaches can support rural development.*

## Recommendations for Ethical Biofuel Supply Chains

- **Land Tenure Protections:** Secure land rights for indigenous and smallholder communities.

- **Sustainability Standards:** Expand certification schemes to include social impact audits.

- **Inclusive Policy Design:** Involve local communities in project planning and benefit sharing.

- **Zoning and Safeguards:** Implement land-use planning to prevent food and water conflicts.

**Ethical Risk Zones in Biofuel Supply Chains**

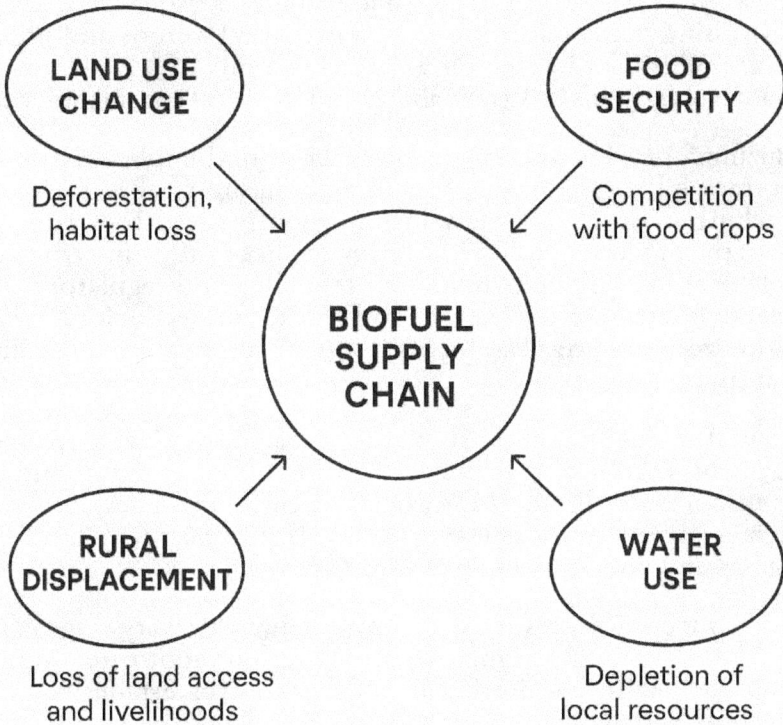

LAND USE CHANGE

FOOD SECURITY

Deforestation, habitat loss

Competition with food crops

BIOFUEL SUPPLY CHAIN

RURAL DISPLACEMENT

WATER USE

Loss of land access and livelihoods

Depletion of local resources

*\* Conceptual diagram highlighting hotspots for ethical risks: land use, water stress, biodiversity loss and food system interference.*

## Sustainable Feedstock Sourcing and Supply Chain Optimization

Sustainable feedstock is crucial for the success of biofuel production, requiring thoughtful consideration of environmental impacts, costs and supply reliability. Developing dependable feedstock sources that adhere to environmental standards poses challenges and opportunities for innovation.

To achieve sustainable feedstock sourcing, it is essential to establish diverse supply networks capable of enduring market changes and meeting stringent environmental standards. This entails strategically choosing feedstock varieties based on regional availability, seasonal fluctuations and processing needs. Utilizing farm waste, energy crops and organic waste, as mentioned in chapter 1, presents a mix of advantages and disadvantages in terms of sustainability and logistical considerations.

Optimizing feedstock sourcing involves the critical implementation of robust traceability systems that benefit modern supply chains with digital tracking solutions. These solutions monitor feedstock from the source to the processing facility, ensuring quality control and providing valuable data for emissions reduction reporting and regulatory compliance. By integrating IoT sensors and blockchain technology, we can revolutionize how we track and verify feedstock sustainability credentials in the future.

Transportation logistics also play a crucial role in sustainable feedstock management. Establishing regional collection hubs can significantly reduce transportation distances and associated emissions. These hubs serve as intermediary points for assessing feedstock quality and initiating processing, streamlining the supply chain and reducing waste.

Quality control measures must be implemented at every stage of the supply chain, including regular feedstock testing, moisture monitoring, real-time quality monitoring and quality control. Establishing clear specifications and maintaining strong supplier relationships helps ensure reliable feedstock quality while supporting sustainable farming practices.

**Key elements of sustainable feedstock sourcing:**

- Regional sourcing strategies to minimize transportation impact

- Diversification of feedstock sources for supply security

- Implementation of digital tracking and verification systems

- Development of quality control protocols

- Establishment of fair pricing mechanisms for suppliers

To ensure sustainable feedstock, it is crucial to collaborate with farmers and implement regenerative agriculture practices. These partnerships should prioritize capacity building, knowledge sharing and creating mutual value. Organizations can support the quality and continuation of their feedstock supply and rural economic development by investing in supplier development programs.

A comprehensive strategy for securing feedstock should take into account climate impacts, market volatility and regulatory changes. Implementing contingency plans and maintaining buffer stocks can help mitigate supply chain disruptions and ensure uninterrupted operations.

Supply chain optimization goes beyond operational efficiency to include environmental and social considerations. This holistic approach necessitates regular monitoring of supply chain performance, focusing on environmental metrics such as carbon footprint, water usage and social impact. Conducting regular audits and implementing continuous improvement programs can help identify opportunities for enhanced sustainability and operational efficiency.

**Essential optimization strategies include:**

- Regular supply chain performance assessments

- Implementation of sustainability metrics and reporting

- Development of supplier capacity-building programs

- Integration of digital monitoring and control systems

- Establishment of circular economy principles

By leveraging artificial intelligence and machine learning algorithms, companies can optimize supply chain operations, predict market fluctuations and proactively identify potential disruptions. Organizations must remain flexible and adaptable, being prepared to embrace emerging technologies and practices while staying committed to emission reduction goals..

# Infrastructure Development for Biofuel Storage and Distribution

Establishing a resilient biofuel storage and distribution infrastructure is essential for effectively shifting towards sustainable energy systems. One of the complexities in biofuel storage arises from the unique chemical properties of the fuel. Unlike traditional fossil fuels, numerous biofuels are hygroscopic, indicating their ability to absorb moisture from the air. Improper storage can result in quality deterioration. Therefore, specialized storage facilities equipped with moisture control mechanisms and materials resistant to corrosion are necessary to mitigate this issue.

**Essential storage infrastructure requirements include:**

- Temperature-controlled storage tanks

- Moisture monitoring systems

- Corrosion-resistant materials and coatings

- Emergency containment systems

- Regular maintenance protocols

Key considerations for efficient delivery while maintaining fuel quality include pipeline compatibility, transportation logistics and terminal facilities. Integrating biofuel distribution systems with existing fuel infrastructure requires careful engineering and strategic planning to minimize disruption while maximizing efficiency.

One of the most significant challenges in infrastructure development is compatibility with various biofuel blends. Storage and distribution systems must be designed to handle different fuel specifications while maintaining strict quality control. Addressing this challenge often requires implementing advanced monitoring systems and flexible storage solutions that adapt to changing fuel compositions.

Terminal facilities play a crucial role in the biofuel distribution network, as they must be equipped with specialized blending capabilities, quality testing laboratories and efficient loading/unloading systems.

When developing biofuel infrastructure, safety considerations are of utmost importance. This includes equipping storage facilities with advanced fire suppression systems, vapor recovery units and emergency response capabilities. Additionally, regular safety audits and maintenance programs are crucial for ensuring the integrity of storage and distribution systems, as well as for protecting personnel and the environment.

**Key safety infrastructure components include:**

- Fire detection and suppression systems

- Vapor recovery units

- Emergency shutdown systems

- Spill containment measures

- Environmental monitoring equipment

When considering cost implications in infrastructure development, a careful balance between initial investment and long-term operational efficiency is essential. While the upfront costs of specialized storage and distribution systems may be significant, strategic planning and phased implementation can help manage the financial impact while ensuring system reliability.

Storage facilities need to be adequately designed to protect against climate impact and extreme weather events. By implementing backup systems and redundancy measures, continuous operations can be ensured during adverse conditions.

To remain compliant and to address future needs, a successful biofuel infrastructure requires close collaboration between various stakeholders, including regulators, equipment manufacturers and end-users.

Key considerations should involve designing flexible systems that can adapt to new fuel specifications and increased capacity requirements. By integrating sustainable design principles and energy-efficient operations, long-term environmental and economic viability can be ensured. Regular inspections, preventive maintenance and performance monitoring are crucial for identifying potential issues before they disrupt operations. This proactive approach to infrastructure management helps to minimize downtime while maximizing system efficiency.

## Adapting and Expanding Biofuel Infrastructure: Technical, Logistical, and Economic Realities

### Infrastructure-Specific Technical Challenges

Biofuels present diverse handling challenges depending on their chemical composition. Below is a comparison of key infrastructure compatibility factors by fuel type:

| Biofuel Type | Storage/Handling Needs | Compatibility Issues |
|---|---|---|
| Ethanol | Sealed tanks, lined pipelines | Hygroscopic; not pipeline-compatible |
| Biodiesel (B20) | Heated tanks in cold climates | Gelling risk, microbial growth on tank bottoms |
| Renewable Diesel | Drop-in compatible with diesel infrastructure | Minimal — compatible with pipelines, terminals |
| SAF (Jet Fuel) | Aviation-certified blending and filtration | Requires separate storage to avoid contamination |
| Biogas (RNG) | Pressurized or liquefied systems | Needs compression, odorization, or liquefaction |

## Cost Benchmarks and Logistical Complexity Overview

Estimated infrastructure upgrade and development costs by fuel type and application.

| Component | Fuel Type | Upgrade Cost Range (USD) | Key Cost Drivers |
|---|---|---|---|
| Tank lining and heating | Ethanol, Biodiesel | $150K–$800K per tank | Moisture/temp control |
| Terminal blending equipment | All | $400K–$2M | Inline blending, filtration |
| Dedicated SAF infrastructure | SAF | $1M–$6M | Jet-grade tanks, hydrant systems |
| Pipeline corrosion retrofits | Ethanol, Biodiesel | $5M+ per 10 miles | Coating, sensors |
| Modular pre-processing hub | Biomass | $2M–$5M | Mobile logistics, remote processing |

### Case Example Infrastructure
***SAF Blending at Schiphol Airport (Netherlands):***
*Schiphol integrated a SAF blending unit into its hydrant fueling system. It included jet-grade isolation tanks and metered in-line blending. This reduced fuel carbon handling emissions by 80%.*

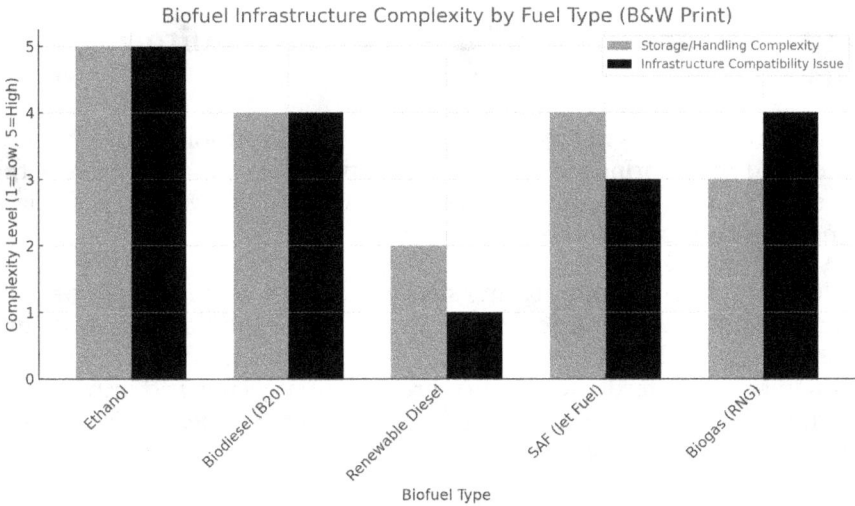

Biofuel Infrastructure Complexity by Fuel Type (B&W Print)

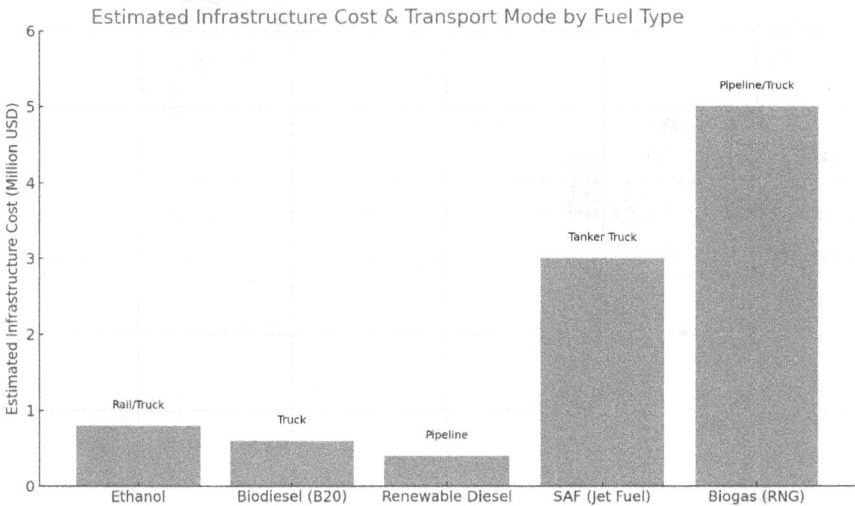

Estimated Infrastructure Cost & Transport Mode by Fuel Type

## Conclusion

Infrastructure is a foundational element of successful biofuel transitions. By addressing technical, logistical, and financial barriers proactively, stakeholders can create resilient, scalable, and low-emission energy networks.

# Digital Integration and Supply Chain Monitoring Systems

The integration of digital technologies into biofuel supply chains has revolutionized monitoring, managing and optimizing operations across the entire value chain, dramatically improving supply chain visibility, efficiency and emission reduction outcomes.

Modern supply chain monitoring systems leverage advanced sensors, Internet of Things (IoT) devices and real-time data analytics to provide unprecedented visibility into operations, enabling continuous tracking of key parameters such as feedstock quality, processing efficiency and distribution logistics. The ability to monitor these metrics in real time allows for rapid response to potential issues and optimization of resource utilization.

Digital integration in biofuel supply chains extends beyond basic monitoring to include predictive analytics and automated decision-making systems, helping organizations anticipate potential disruptions, optimize inventory levels and maintain consistent product quality. Implementing artificial intelligence and machine learning algorithms has enabled more sophisticated analysis of supply chain data, leading to improved forecasting accuracy and operational efficiency.

## Key components of digital monitoring systems include:

- Real-time sensor networks for quality control

- Automated data collection and analysis platforms

- Predictive maintenance systems

- Supply chain visibility dashboards

- Environmental impact tracking tools

> "It's not enough to spot risks—we need procurement strategies that can absorb them. Our contracts, our logistics, our people... all of it has to flex with the fuel."
>
> — Luis Ortega, Director of Procurement, Global Mining Group

The integration of blockchain technology has emerged as a powerful tool for ensuring transparency and traceability in biofuel supply chains. By creating an immutable record of transactions and quality certifications, blockchain systems help verify emission reduction claims and maintain regulatory compliance. This technology has proven particularly valuable in managing complex international supply chains where multiple stakeholders require access to reliable information.

One of the most significant advantages of digital integration is the ability to optimize logistics and transportation networks. Advanced routing algorithms and real-time tracking systems help reduce transportation costs while minimizing environmental impact, as these systems can dynamically adjust to changing conditions, ensuring efficient delivery while maintaining product quality.

Data security and system reliability are crucial considerations in digital integration. Supply chain monitoring systems must be designed with robust cybersecurity measures and redundancy protocols to protect sensitive information and ensure continuous operation. Regular system audits and updates help maintain security while adapting to emerging threats and technological advances.

To successfully implement digital monitoring systems, staff members need to be equipped with the necessary skills to utilize these systems effectively and interpret the data they provide. Developing comprehensive training programs and establishing clear protocols for system usage are essential steps in this process.

Another critical aspect of digital transformation in biofuel supply chains is integrating existing enterprise systems. Modern monitoring systems should be able to communicate with enterprise resource planning (ERP) systems, quality management systems and other business applications to provide a comprehensive view of operations. This integration not only helps ensure data consistency but also enables more effective organizational decision-making.

**Essential considerations for system integration include:**

- Data standardization protocols

- Interface compatibility

- System scalability requirements

- Performance monitoring metrics

- User access controls

The implementation of digital monitoring systems has enabled more sophisticated quality control and compliance management approaches. Automated sampling, testing procedures, and real-time data analysis help ensure consistent product quality while reducing non-compliance risk. These systems can automatically flag potential issues and initiate corrective actions before problems escalate.

Moving forward, the continued evolution of digital technologies promises even greater capabilities in supply chain monitoring and optimization. Developing advanced analytics tools, improved sensor technologies, and more sophisticated automation systems will further enhance our ability to manage complex biofuel supply chains efficiently and sustainably.

When considering cost implications in digital integration, it is essential to balance the initial investment against long-term operational benefits. While implementing comprehensive monitoring systems requires significant resources, the resulting efficiency, quality control and risk management improvements often justify the investment. Organizations should develop clear ROI metrics to evaluate system performance and guide future investments in digital capabilities.

Roles and responsibilities for system management need to be defined, data quality standards established and proper controls for system access and modification implemented. Regular reviews of system performance and user feedback are essential to ensure that digital tools continue to effectively meet organizational needs.As we conclude our exploration of sustainable supply chain development in the biofuel sector, it is evident that the success of any biofuel operation depends on the careful orchestration of infrastructure, digital integration and resource management strategies. Throughout this chapter, we have discussed how transforming traditional supply chains into sustainable, technology-enabled systems requires a delicate balance between operational efficiency and environmental stewardship.

The implementation of blockchain-based tracking systems and advanced monitoring technologies has shown the transformative potential of digital integration in supply chain management, enhancing transparency, traceability and enabling real-time optimization of resource utilization. Furthermore, establishing local collection hubs and forming strategic partnerships with suppliers has proven critical in building resilient supply networks capable of withstanding market volatility while upholding sustainable practices.

Developing infrastructure for biofuel operations is a critical challenge that requires specialized storage facilities, distribution networks and quality control systems, all of which demand significant investment and careful planning. In this context, integrating digital technologies and sustainable design principles has been shown in various case examples to optimize investments and ensure long-term operational viability.

Supply chain monitoring systems play a crucial role in maintaining product quality and regulatory compliance, leveraging advanced sensor networks, predictive analytics, and automated quality control systems to revolutionize the management of complex biofuel supply chains. These technologies not only enhance operational efficiency but also provide valuable data for continuous improvement and emission reduction reporting.

Looking ahead, as technological advancements and regulatory requirements continue to evolve, they are likely to shape the future of sustainable supply chain development in the biofuel sector. Companies that prioritize flexible, scalable and sustainable infrastructure will be at the forefront of leading the way in renewable energy.

## Key Takeaways

- Blockchain, IoT, and local sourcing partnerships dramatically improve transparency, traceability and resilience.

- The need for specialized infrastructure development and quality control systems

- Sustainable supply chains require decentralized infrastructure ( e.g., regional hubs, modular storage) and robust digital tracking systems.

- Regional collection and preprocessing reduce emissions costs and bottlenecks, especially in remote or resource-limited environments.

As we transition to the next chapter on investment analysis, we will delve into how these supply chain considerations intersect with financial planning and market opportunities in the biofuel sector. Building a solid foundation of sustainable supply chain development is crucial for understanding the economic impacts and investment potential of biofuel projects.

Chapter 4

# INVESTMENT ANALYSIS

*Financial Models and Market Opportunities in Biofuels*

This chapter is designed for financial analysts, investors, project developers and strategic planners evaluating the commercial viability of biofuel ventures. While the technical underpinnings are essential, our focus here is on monetization pathways, risk mitigation tools and investment decision frameworks across global markets.

The landscape of biofuel investment is marked by a complex interplay between traditional financial metrics and evolving Greenhouse Gas (GHG) emissions criteria. Investors are required to master new valuation frameworks that encompass both monetary and environmental returns. As capital markets increasingly acknowledge the strategic significance of renewable energy transitions, understanding the nuances of biofuel project finance has become essential for investors aiming to capitalize on this growing sector while effectively managing associated risks.

Within this dynamic investment environment, traditional assessment metrics are being reshaped by evolving sustainability imperatives and regulatory frameworks. The integration of environmental, social and governance (ESG) criteria has become increasingly crucial in evaluating biofuel investments, necessitating a more nuanced approach to project valuation and risk assessment. Successful investment strategies in the biofuel sector hinge on a deep understanding of conventional financial metrics and emerging value drivers specific to renewable energy projects.

Key considerations include recognizing the potential impact of carbon pricing mechanisms, renewable energy credits and government incentives on project economics. The interplay of these factors creates a complex decision-making environment where meticulous analysis and strategic foresight are paramount.

**Case Example**
*An illustration of this complexity arose during the development of a biofuel facility designed to convert agricultural waste into renewable diesel.*
*Traditional valuation methods fell short in capturing the full scope of the project's potential, prompting the creation of a hybrid valuation model that integrated conventional financial metrics with sustainability factors. This innovative approach encompassed carbon credit revenues, regulatory incentives, potential policy changes and traditional cash flow projections.*

The pivotal insight gained from this experience recognized that a project's true value often extends beyond immediate returns, particularly in the rapidly evolving biofuel sector. Strategic positioning for future market opportunities became as crucial as current profitability metrics, reshaping the fundamental approach to biofuel investment analysis. This revelation emphasized the importance of evaluating a broader range of value drivers in renewable energy projects.

The model developed through this process has since become a foundational tool for future investment evaluations, enabling stakeholders to make more informed decisions in the biofuel sector. It demonstrates how integrating emission metrics with traditional financial analysis can create a more comprehensive and accurate picture of investment potential in renewable energy projects.

Delving deeper into investment analysis in the biofuel sector, we will explore sophisticated financial modeling techniques, risk assessment frameworks and market valuation methods that reflect the sector's unique characteristics. Understanding these tools and methodologies is essential for investors, project developers and industry stakeholders seeking to capitalize on the growing opportunities in sustainable energy while effectively managing associated risks.

## Financial Modeling and Valuation Metrics for Biofuel Projects

To effectively model and value biofuel projects, a comprehensive framework is essential. This framework should encompass both traditional financial metrics and emerging value drivers unique to renewable energy investments. Given the complexity of these projects, sophisticated analysis tools are necessary to consider various variables. These variables range from feedstock costs and processing efficiency to regulatory incentives and environmental credits.

**Key financial metrics for biofuel project evaluation include:**

- Net Present Value (NPV) and Internal Rate of Return (IRR) adjusted for renewable energy credits

- Levelized Cost of Fuel Production (LCFP)

- Carbon credit revenue potential

- Government incentive capture rates

- Working capital requirements for feedstock management

While the traditional discounted cash flow (DCF) model is fundamental, adjustments are necessary for biofuel-specific value drivers. These adjustments should involve modifying discount rates to reflect the unique risk profile of biofuel projects and incorporating sensitivity analysis for key variables like feedstock prices and regulatory changes.

**Example: Financial Scenarios – 50 MGPY (million gallons per year) Renewable Diesel Plant**

| Metric | Base Case | Optimistic Case | Conservative Case |
|---|---|---|---|
| Capital Cost | $300 million | $280 million | $320 million |
| Annual Output | 50 million gallons | 55 million gallons | 45 million gallons |
| Feedstock (waste oil) cost ($/gal) | $1.50 | $1.25 | $1.75 |
| LCFS + RIN credits ($/gal) | $1.00 | $1.30 | $0.70 |
| Net selling price ($/gal) | $3.50 | $3.80 | $3.20 |
| Operating Cost ($/gal) | $2.20 | $2.00 | $2.40 |
| Net Margin ($/gal) | $1.30 | $1.80 | $0.80 |
| IRR (10-year) | 14% | 20% | 8% |
| NPV (@8% discount) | $85 million | $160 million | $30 million |

*\* This analysis shows that even under conservative market assumptions, the plant remains profitable, although with reduced returns. In jurisdictions with strong policy incentives (like California's LCFS), even modest tech performance can yield competitive financial outcomes.*

**Global Insight:**
*In Brazil, IRR on sugarcane ethanol is driven by RenovaBio carbon credits and stable feedstock costs, whereas in Germany, projects often hinge on RED II compliance and long-term EU subsidies.*

It is crucial to measure efficiency, such as conversion rates and energy use, to accurately assess biofuel projects. Understanding these technical parameters is essential for precise financial projections and risk assessment, especially when evaluating different technology options or scaling decisions.

Risk-adjusted valuation frameworks need to consider both market and technical risks specific to biofuel projects, including feedstock price volatility, technology performance risks and regulatory compliance costs.

Sophisticated modeling approaches often utilize Monte Carlo simulations to analyze the impact of multiple risk factors simultaneously. Working capital management is particularly important in biofuel project modeling due to the seasonal nature of many feedstock supplies, storage requirements and processing schedules that create unique cash flow patterns. An in-depth analysis of inventory holding costs, storage infrastructure requirements and supply chain financing needs is vital for effective financial planning.

Furthermore, environmental credits and regulatory incentives can significantly affect project economics and should be meticulously modeled.

**This includes:**

- LCFS (Low Carbon Fuel Standard) volatility, as in California or British Columbia

- RIN markets – under the U.S. Renewable Fuel Standard

- EU ETS extensions – for SAF and maritime fuels

- Voluntary carbon markets – increasingly relevant for aviation and high-blend projects

- Tax incentives and subsidies

- Compliance cost projections

**Case Example**
*A U.S. cellulosic ethanol project hedged its exposure by entering long-term offtake agreements indexed to both Brent crude and LCFS floors. This stabilized cash flow enables access to green bond financing with ESG-linked performance metrics.*

In addition to financial metrics, the valuation process must also consider the strategic value of biofuel projects beyond purely financial returns. Key factors such as corporate emission reduction goals, supply chain resilience and competitive positioning in an increasingly carbon-conscious market environment. In many cases, these strategic considerations often justify investment decisions that might appear marginal, based on traditional financial metrics alone.

Furthermore, operational flexibility and scalability should be incorporated into financial models by leveraging accurate options analysis. Adopting this approach helps quantify the value of strategic alternatives, such as capacity expansion, feedstock switching, or technology upgrades. Ultimately, the ability to adapt to changing market conditions and technological developments can significantly enhance project value.

## Levelized Cost of Fuel (LCOF) by Technology (2024 estimates)

| Biofuel Type | Feedstock | LCOF ($/gallon) | Fossil Equivalent ($/gallon) | Key Notes |
|---|---|---|---|---|
| Corn Ethanol (US Midwest) | Corn | $1.30–$1.60 | Gasoline: $2.90 | Mature, vulnerable to corn/feed price swings |
| Cellulosic Ethanol | Ag residue/switch grass | $2.50–$3.50 | Gasoline: $2.90 | High capex; upside in waste valorization; tech still scaling |
| Algae Biodiesel | Photobioreactor algae | $4.00–$6.00 | Diesel: $3.25 | High yield potential, early-stage; cost decline expected with tech maturity |
| HEFA Renewable Diesel | Waste oils/fats | $2.20–$3.00 | Diesel: $3.25 | Commercial scale; attractive with LCFS and RIN credits |
| Gas Fermentation Ethanol | Industrial CO/$CO_2$ waste gases | $2.70–$3.50 | Gasoline: $2.90 | Proven tech (e.g., LanzaTech); reliant on industrial partnerships |
| SAF (Alcohol-to-Jet, HEFA) | Ethanol / Fats | $3.00–$5.00 | Jet-A: $2.80 | Fastest-growing sector, driven by airline offtake and mandates |

**Takeaway:** Biofuel cost competitiveness is improving rapidly. Policy-driven credits (RINs, LCFS, SAF tax credits) can push many fuels below parity with fossil counterparts in strategic markets.

Additionally, benchmarking against comparable projects and industry standards provides an essential context for valuation assumptions. However, regional variations in feedstock availability, regulatory environments and market conditions must be considered. Using standardized metrics enables meaningful comparisons while still recognizing project-specific factors that influence performance.

## International Funding Mechanisms and Market Access

Investment analysis must consider jurisdictional differences:

- **Asia:** Japanese and Korean SAF demand is growing, backed by fuel import credits

- **EU:** Horizon Europe and Innovation Fund grants enable pilot-to-commercial scale transitions

- **Africa:** Green hydrogen-ammonia hubs (e.g., Namibia) are emerging biofuel-adjacent opportunities

- **South America:** Brazil's BSBios and Raízen leverage agricultural waste with strong domestic demand

- **Australia:** Future Made in Australia Innovation Fund grant funding support for renewable and low-emission technologies

**Biofuel Investment Ecosystem**

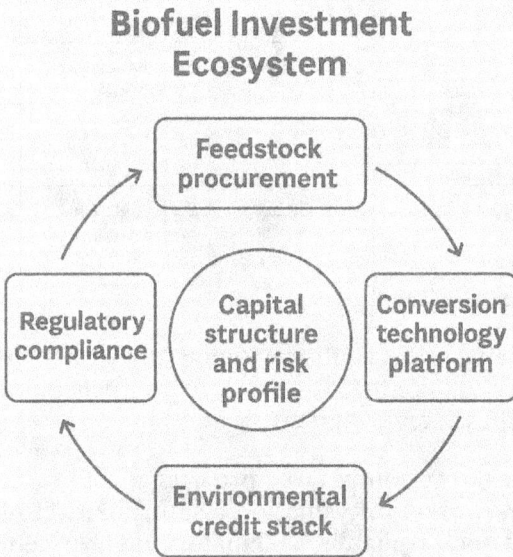

**Biofuel Investment Ecosystem**

- Feedstock procurement
- Conversion technology platform
- Capital structure and risk profile
- Regulatory compliance
- Environmental credit stack

To strengthen financial assessments, financial models should incorporate scenario analysis to evaluate project resilience under different market conditions. This method involves stress testing assumptions and proactively developing contingency plans. By demonstrating strong project finances across various scenarios, organizations can attract investors and secure funding with greater confidence.

## Risk Assessment and Mitigation Strategies in Biofuel Investments

To ensure success, smart biofuel investment needs careful risk assessment along with strong mitigation plans. Given that the biofuel sector is highly unpredictable, it presents various challenges that demand thorough evaluation. In particular, organizations must carefully consider technical, operational, market and regulatory uncertainties to develop effective risk management strategies.

**Key risk categories that require systematic assessment include:**

- Feedstock supply volatility and price fluctuations

- Technology performance and scaling challenges

- Regulatory compliance and policy changes

- Market demand uncertainty and competition

- Environmental and sustainability compliance risks

- Capital access

To initiate effective risk assessment, a thorough due diligence process is crucial, examining both quantitative and qualitative risk factors. This process involves analyzing feedstock, conducting technology testing and researching the market to identify potential risks and opportunities. It is important to not only focus on direct operational risks but also consider broader systemic risks that could impact the project's viability.

## Biofuel Investment Risk Matrix

| Risk Category | Likelihood | Impact | Example | Suggested Mitigation |
|---|---|---|---|---|
| Feedstock price volatility | High | High | Palm oil price swings affect HEFA margins | Long-term procurement contracts; feedstock diversification |
| Regulatory/ policy change | Medium | High | RIN value collapse or LCFS adjustment | Policy hedging via multi-region deployment; regulatory buffers |
| Technology performance | Medium | High | Microbial yield failure in cellulosic ethanol process | Pilot validation, staged scaling, performance guarantees from vendors |
| Market demand shifts | Medium | Medium | Aviation SAF mandates delayed or weakened | Diverse offtake agreements across sectors (aviation, marine, trucking) |
| Operational risks | Low | Medium | Enzyme deactivation or downtime in fermentation reactors | Preventive maintenance, redundancy systems |
| Financial/ capital access | Medium | Medium | Delayed equity round or debt financing interruption | Credit lines; staged fundraising; blended finance (green bonds + equity) |
| Sustainability reputation | Low | High | NGO campaigns over feedstock (e.g., deforestation concerns with palm oil) | Traceability systems; certification (e.g. RSB, ISCC) |

**Takeaway:** Investors must actively manage not only technical and financial risks, but also emerging ESG and policy uncertainties. A robust, pre-modeled mitigation strategy boosts bankability and investor confidence.

A structured approach to risk mitigation is essential, which includes developing robust contingency plans and establishing clear risk monitoring and response protocols. Key strategies involve implementing sophisticated feedstock procurement strategies and maintaining operational flexibility through diversified technology platforms.

Additionally, building strong relationships with key stakeholders along the value chain is imperative. In biofuel investments, operational risk management demands special attention. Key measures include implementing rigorous quality control systems, ensuring adequate insurance coverage, and developing comprehensive maintenance programs. Regular performance monitoring and preventive maintenance schedules are critical in identifying potential issues before they escalate into significant problems.

Mitigating market risk typically requires utilizing a mix of strategic and financial tools. For instance, long-term offtake agreements can offer revenue stability and hedging strategies are effective in managing feedstock price volatility. Another approach is diversifying across various market segments or geographic regions to minimize exposure to market-specific risks.

Managing regulatory risk effectively involves nurturing strong relationships, staying informed on policy updates and being adaptable. Essential strategies include incorporating compliance buffers into operational plans and openly communicating with stakeholders about regulatory issues and responses.

**Technology risk mitigation strategies should focus on:**

- Thorough technology validation through pilot testing

- Staged implementation approaches for new processes

- Strong warranties and performance guarantees from technology providers

- Ongoing research and development investment

- Regular technology assessment and upgrade planning

Financial risk management encompasses more than just traditional hedging strategies. It also involves carefully structuring project financing, maintaining adequate working capital reserves and developing contingency funding plans. Strategies may include establishing revolving credit facilities or cultivating relationships with multiple funding sources to ensure financial stability.

In biofuel projects, special attention must be given to environmental and sustainability risks. Successful management in this area entails implementing robust environmental programs, fostering strong community relationships and providing transparent reporting of results. Environmental audits play a crucial role in maintaining compliance and identifying issues early on.

Effective risk mitigation also necessitates the establishment of robust governance structures and the maintenance of clear communication channels among project stakeholders. Regular risk reviews and updates to mitigation strategies are essential to ensure that risk management remains adaptable and responsive to changing conditions.

By maintaining a regularly updated risk register, project risks can be effectively managed. This entails key components such as assigning clear risk ownership, defining specific mitigation actions and consistently monitoring risk indicators.

## Market Risk Management Strategies and Financial Hedging

Biofuel markets are exposed to price swings in feedstock, oil, and credit markets (e.g., RINs, LCFS). Tools to manage volatility include:

- Long-term offtake agreements indexed to credit values

- Vertical integration into feedstock or distribution

- Financial hedging instruments for oil and grain futures

### Case Example
*A California-based renewable diesel firm signed 10-year contracts with municipal bus fleets, indexed to LCFS credit floors, stabilizing cash flows during credit price downturns. Ultimately, successful risk management in biofuel investments necessitates a balanced approach that acknowledges both the potential rewards and risks inherent in such projects. Proactive risk management and adaptability are essential factors for achieving success in this regard.*

# Market Entry Strategies and Funding Mechanisms for Biofuel Ventures

Achieving successful market entry in the biofuel sector requires a carefully orchestrated approach that seamlessly combines strategic planning with appropriate funding. Given the complexity of the biofuel market, a thorough understanding of both technical requirements and financial structures is necessary to ensure sustainable growth and market penetration.

**Key market entry strategies for biofuel ventures include:**

- Strategic partnerships with established industry players

- Phased market penetration focusing on specific sectors or regions

- Vertical integration opportunities across the supply chain

- Technology licensing and intellectual property strategies

- Joint venture arrangements for risk-sharing

The selection of an appropriate market entry strategy depends heavily on factors such as available resources, technological maturity and target market characteristics. A step-by-step approach is often best, minimizing risk exposure while building skills and trust.

Funding mechanisms for biofuel ventures have evolved significantly, reflecting the industry's maturing nature and increasing investor confidence in renewable energy projects. Biofuel projects benefit from innovative funding that complements traditional methods.

**Common funding mechanisms include:**

- Green bonds and sustainability-linked loans

- Public-private partnerships (PPPs)

- Government grants and incentive programs

- Strategic corporate investment

- Venture capital and private equity funding

The selection of appropriate funding mechanisms must align with the project scale, development stage and risk profile. For early-stage ventures, a combination of government support and private investment can be beneficial. In contrast, more mature projects have access to a broader range of financing options.

**Case Example**
***Fulcrum BioEnergy's Sierra BioFuels Plant*** *in Nevada is among the first U.S. commercial facilities converting municipal solid waste (MSW) into sustainable aviation fuel (SAF). Fulcrum's success lies in its multi-pronged investment and risk mitigation approach.*

## Key Strategy:

| Category | Fulcrum Approach |
|---|---|
| Technology Risk | Validated via a multi-year pilot and 10-year phased roadmap |
| Offtake Security | 10-year SAF supply contract with United Airlines |
| Funding | $105 million DOE loan guarantee; private equity + bond issue |
| Policy Support | LCFS credits (California), RINs, and SAF blending mandates |
| Partnerships | BP (fuel logistics), United Airlines, Cathay Pacific |

**Outcome:** Despite high capex (~$300M), Fulcrum's project is expected to break even in under 8 years, mainly due to guaranteed revenue from long-term offtake and carbon credits.

**Takeaway:** Fulcrum demonstrates how biofuel developers can de-risk early-stage investments through government support, strategic partnerships and offtake contracts—making advanced tech bankable.

Project developers must carefully consider the implications of different funding structures on operational flexibility and long-term strategic objectives, as the terms and conditions of various funding sources can significantly impact project economics and operational decision-making capacity.

Having strong business plans is crucial for securing funding, as they should demonstrate market understanding, realistic finances and effective risk management. Investors are increasingly interested in projects that offer strong financial returns, measurable environmental benefits and clear competitive advantages.

Government support, such as policy frameworks like renewable fuel standards, tax incentives and carbon pricing mechanisms, plays a critical role in market entry. These policies significantly influence market dynamics and project viability. Therefore, successful market entry strategies must take these policy drivers into account while also maintaining flexibility to adapt to regulatory changes.

Infrastructure development is another crucial consideration in market entry planning. The availability of storage, transportation and distribution infrastructure can significantly impact project economics and market access. New market entrants often find cost-effective solutions through strategic partnerships with existing infrastructure operators.

Risk mitigation strategies are essential for attracting funding and ensuring successful market entry. These strategies involve developing robust feedstock supply agreements, securing offtake contracts and implementing comprehensive insurance coverage. Access to preferred funding sources often depends on the ability to demonstrate effective risk management.

Technology validation and scale-up strategies also require particular attention in market entry planning. Pilot projects and demonstration facilities help build investor confidence and provide valuable operational experience. Staged technology deployment is often more attractive to investors than aggressive scaling plans.

Market positioning and competitive analysis play a crucial role in informing both entry strategies and funding approaches. By understanding market segments, competitor capabilities and customer needs, one can identify viable market opportunities and appropriate positioning strategies. It is important to consider both current market dynamics and potential future developments in this analysis.

Additionally, successful market entry requires careful attention to timing and market conditions, as economic cycles, policy changes and technological developments can significantly impact project viability. The ability to adjust entry strategies and funding approaches in response to changing conditions often determines long-term success.

Environmental, Social and Governance (ESG) considerations are increasingly influencing both market entry strategies and funding availability. Developing strategic partnerships emerges as a crucial element in successful market entry, providing access to essential resources, market knowledge and distribution channels while helping manage risk and capital requirements. Partner selection criteria should align with long-term strategic objectives and operational requirements.

As we conclude our examination of investment analysis in the biofuel sector, several critical insights emerge that shape the path forward for investors, project developers and industry stakeholders. Evaluating biofuel investments now considers both financial profits and environmental sustainability, reflecting the maturing nature of the industry and the increasing sophistication of investment approaches.

Developing hybrid valuation models that integrate environmental impact, regulatory compliance, technological innovation and conventional financial metrics has proven essential for accurate project assessment. These models provide a more comprehensive framework for decision-making, enabling investors to capture the full value potential of biofuel projects while effectively managing associated risks.

> "Investors don't just want clean returns—they want credible roadmaps. ESG alone won't secure capital unless the fundamentals hold up."
> — Anya Becker, ESG Investor & Former Energy Market Analyst

Analyzing financial models and risks has highlighted the need for a systematic project evaluation approach. The success of biofuel investments increasingly depends on the ability to navigate complex market dynamics, maintain operational flexibility and adapt to evolving regulatory requirements.

The case example of the agricultural waste conversion facility exemplifies how innovative valuation approaches can unlock hidden value in biofuel projects. By incorporating carbon credit revenues, regulatory incentives and strategic positioning alongside traditional metrics, investors can better understand and capture the full potential of these investments. This comprehensive approach to valuation has become increasingly critical as the industry continues to mature and evolve.

Investments in biofuels require improved strategies and valuation methods to succeed. By integrating financial and emission data, better decisions can be made and risks managed more effectively.

Analyzing funding mechanisms and market entry strategies reveals the importance of aligning financing structures with project characteristics and development stages. Diverse funding options, such as green bonds and strategic partnerships, offer flexibility in structuring investments and managing risk exposure.

The future of biofuel investment is closely tied to technological advancements, government policies and market trends. Sustainable energy investment necessitates a balanced consideration of its opportunities and challenges. The frameworks and methodologies outlined in this chapter establish a solid foundation for navigating the intricate yet promising market. Ultimately, successful biofuel investment hinges on combining thorough financial analysis with forward-thinking sustainability and environmental impact strategies. As the industry matures, those who can seamlessly integrate these elements while upholding operational excellence and strategic adaptability will be best positioned to seize the growing opportunities in the biofuel sector.

---

*■ Reality Check: Policy Shifts Can Undermine Financial Models*

*Many biofuel ventures hinge on favorable regulatory conditions, like LCFS credits, SAF tax breaks or carbon pricing. But these incentives are subject to political change. In one recent case, a Southeast Asian biodiesel producer saw margins collapse when export tariffs changed and European regulators delayed emissions certification. Investors must build policy scenarios into project models, not just tech assumptions.*

*Risks to Monitor:*

- *Sudden changes to renewable fuel standards (RFS, RED II)*
- *Over-reliance on one type of environmental credit*
- *Delays in infrastructure-linked funding (e.g., storage, pipelines)*

---

## Final Recommendations

- **For Investors:** Model downside scenarios with and without credit incentives

- **For Developers:** Align CapEx profiles with credit monetization schedules

- **For Policymakers:** Foster credit stability and permit stackability across jurisdictions

# Key Takeaways

- **Biofuel investment analysis requires hybrid financial models** that integrate traditional metrics (e.g., NPV, IRR) with sustainability value drivers such as carbon credits, ESG impacts, and regulatory incentives to capture a project's full long-term value.

- **Risk assessment is multi-dimensional,** encompassing feed-stock volatility, regulatory uncertainty, technology scalability and environmental compliance. Mitigation strategies include diversi-fied supply chains, offtake agreements and proactive stakeholder engagement.

- **Innovative funding mechanisms**—including green bonds, PPPs, and sustainability-linked loans—are increasingly essential for financing biofuel ventures, especially when paired with robust business plans and precise ESG alignment.

- **Strategic market entry depends on partnerships, infra-structure access and regulatory awareness,** with phased roll-outs, licensing deals and joint ventures enabling both scalability and risk-sharing.

- **Scenario planning and sensitivity analysis** are critical tools to ensure financial resilience, helping investors stress-test biofuel projects under fluctuating market conditions and evolving policy frameworks.

# Leave A Review

*Thank You!*

If this book resonated with you or supported you in any way, you're warmly invited to leave a **review** on the platform where you purchased this book — reviews help other readers discover books like this.

If you'd like, you can also access the BIOFUEL READINESS PACK created to support and complement your reading experience.

*Scan Me: REVIEW and access BIOFUEL READINESS PACK*

Chapter 5

# Regulatory Frameworks

*Navigating Global Policy and Compliance Requirements*

As the biofuel industry scales, regulatory frameworks are playing an increasingly central role in shaping market opportunities, guiding investment flows and setting the standards for environmental and social sustainability. This chapter explores the complex ecosystem of global, national and regional regulations that affect biofuel development and outlines how organizations can navigate these frameworks strategically. Understanding this regulatory ecosystem is vital for stakeholders across the value chain, from feedstock suppliers to end-users, as it directly influences operating decisions and investment strategies.

The success of biofuels hinges as much on regulatory navigation as technological innovation. With jurisdictions applying different sustainability standards, tax incentives and compliance mechanisms, organizations must proactively design regulatory strategies from project inception.

Managing cross-border supply chains can pose significant challenges in regulatory navigation.

> **Case Example**
> *A project encountered a complex challenge when navigating international biofuel regulations to establish a cross-border supply chain spanning multiple jurisdictions. Each region maintained distinct regulatory requirements and emission criteria, resulting in initial inefficiencies and compliance gaps when addressed in isolation. A unified compliance matrix addressed overlapping rules and found ways to simplify them, organizing complex, multi-jurisdictional compliance into a manageable framework. The matrix proved invaluable in streamlining the certification processes while ensuring adherence to the most stringent standards across all regions.*

# COMPLIANCE MATRIX

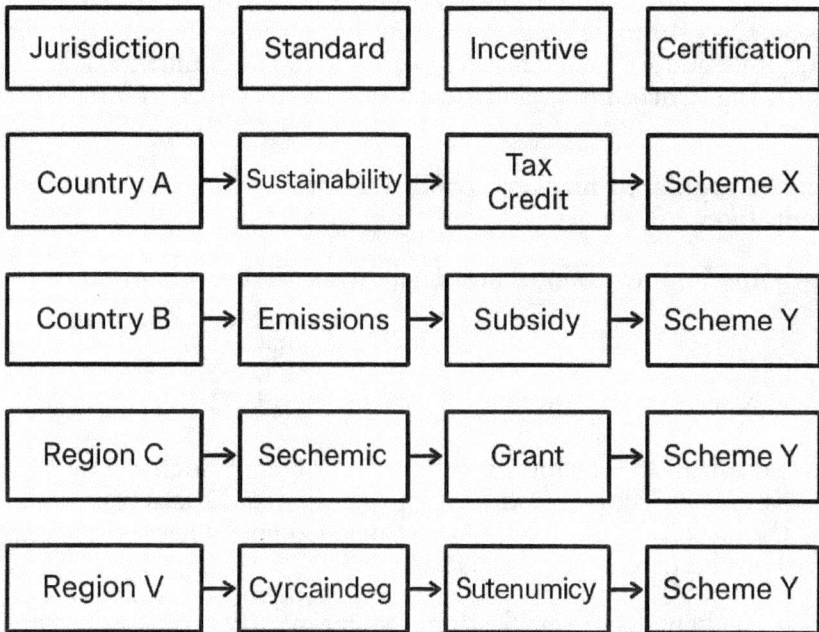

| Jurisdiction | Standard | Incentive | Certification |
|---|---|---|---|
| Country A | Sustainability | Tax Credit | Scheme X |
| Country B | Emissions | Subsidy | Scheme Y |
| Region C | Sechemic | Grant | Scheme Y |
| Region V | Cyrcaindeg | Sutenumicy | Scheme Y |

The key to managing expectations is viewing regulatory compliance not as a burden but as an opportunity to build more resilient and sustainable operations.

Delving deeper, this chapter explores global compliance, risk management and system development with the goal of equipping readers with the tools and understanding necessary to potentially transform regulatory cross-border challenges into competitive advantages in the rapidly evolving biofuel sector.

# Global Compliance Standards and Certification Requirements for Biofuel Production

Global standards and certifications play a crucial role in ensuring the sustainability of biofuels, as they help protect the environment, society and markets. Over the past decade, these standards have undergone significant evolution to address the increasing awareness of environmental issues and the growing demand for standardized quality assurance in international markets.

**To access regulated markets, producers must navigate certification systems like:**

- ISCC (International Sustainability & Carbon Certification)

- RSB (Roundtable on Sustainable Biomaterials)

- National programs (e.g., EPA RFS, Brazil's CBIO credits)

These certification systems assess important factors such as greenhouse gas emissions reduction, biodiversity protection and social responsibility along the entire supply chain, offering detailed guidelines for sustainable biofuel production.

Despite the benefits of certification, the complexity of the requirements can be daunting, especially for organizations new to biofuel production. Balancing compliance with multiple standards while ensuring operational efficiency presents a challenge. To address this, it is advisable to develop an integrated management system that takes a holistic approach to meet certification requirements, rather than treating each one in isolation.

The following key elements typically form the foundation of **global compliance standards:**

- **Greenhouse Gas Emissions Reduction:** Minimum thresholds for lifecycle emission savings

- **Sustainable Feedstock Sourcing:** Requirements for responsible land use and agricultural practices

- **Chain of Custody Documentation:** Tracking systems for verifying sustainable sourcing

- **Social Impact Assessment:** Evaluation of community impacts and labor conditions (refer to chapter 3, Ethical and Community Impacts of Biofuel Supply Chains)

- **Quality Control Parameters:** Technical specifications for biofuel products

To successfully implement these standards, organizations need to establish robust documentation systems, effectively train personnel, regularly audit the process and maintain regular communication with certification bodies to ensure ongoing compliance.

The trend in certifications is moving towards digital verification and blockchain technology. This new technology enhances compliance monitoring, making certification processes easier and more reliable.

The financial implications of certification should not be underestimated. While the initial costs of certification can be substantial, organizations that proactively embrace certification requirements often find themselves in a better position to capitalize on emerging market opportunities and navigate regulatory changes effectively.

Looking ahead, further harmonization of global standards is likely to continue, driven by the need for greater consistency in international markets. Although this trend towards standardization may present challenges in the short term, it will ultimately benefit the industry by reducing compliance complexity and facilitating global trade in sustainable biofuels.

# Policy Mechanisms and Incentive Structures Across Major Markets

Policy mechanisms and incentive structures play a crucial role in shaping the development and adoption of biofuels across major markets. The frameworks vary significantly between regions, creating a complex landscape that industry stakeholders must navigate effectively to optimize their market strategies and investments.

Different countries have adopted diverse policy mechanisms to promote biofuel adoption:

- **Blending Mandates** (e.g., EU RED II, Brazil's RenovaBio)

- **Tax Incentives and Subsidies** (e.g., U.S. Inflation Reduction Act, India's Ethanol Blending Program)

- **Carbon Pricing Schemes** (e.g., California LCFS)

- **Import/Export Tariffs and Sustainability Requirements**

- **Research and Development Support:** Funding programs for advanced biofuel technologies

- **Infrastructure Development Grants:** Financial support for biofuel distribution networks

These tools shape production costs, market access and technology deployment. Policymakers increasingly combine these mechanisms into integrated frameworks that align biofuel development with climate targets and industrial decarbonization goals.

The interaction between different policy mechanisms creates complex market dynamics that necessitate careful analysis. Market responses to policy interventions vary significantly across regions, depending on factors such as existing infrastructure, feedstock availability and industrial base.

For example, regions with robust agricultural sectors often implement policies that leverage local feedstock advantages, while those focused on industrial decarbonization may prioritize policies supporting advanced biofuel technologies.

Effective utilization of policy mechanisms and incentive structures requires a balanced approach that combines technical expertise with strategic planning. Success requires aligning operations with policies, but also adapting to changing markets.

Looking forward, we can expect continued evolution in policy mechanisms as governments adjust their approaches based on market responses and technological developments.

# Evaluating the Effectiveness and Tradeoffs of Biofuel Policy Mechanisms

A comparative analysis of commonly used biofuel policy instruments:

| Policy Mechanism | Strengths | Weaknesses / Risks | Effectiveness Rating |
|---|---|---|---|
| Blending Mandates | Drives demand; provides market certainty | May force uptake of less-efficient fuels; distorts price signals | ★★★★☆ |
| Tax Incentives | Encourages early investment and tech adoption | Risk of subsidy abuse may not drive long-term innovation | ★★★☆☆ |
| Carbon Pricing | Aligns with emissions goals; supports all techs | Politically sensitive; difficult to calibrate | ★★★★☆ |
| LCFS / Credit Systems | Rewards efficiency gains; tech-neutral | Credit volatility can disincentivize long-term investment | ★★★★☆ |
| Biofuel Quotas (EU RED) | Strong policy driver across the EU | Complexity in ILUC tracking; cross-border enforcement | ★★★☆☆ |

*\* Rating based on IEA, ICCT, and World Bank biofuel policy reviews*

## Unintended Consequences and Tradeoffs

Despite good intentions, several biofuel policies have produced unexpected outcomes. U.S. ethanol mandates raised food prices during the 2008 crisis. Southeast Asia's palm biodiesel subsidies led to deforestation, and volatile LCFS credits have caused investment hesitation in emerging markets.

## Political Economy of Biofuel Regulation

Understanding why biofuel policies take the forms they do requires examining the political and economic dynamics driving their development.

Biofuel policy is shaped as much by political economy as by climate science. Agriculture lobbies often influence feedstock-specific subsidies (e.g., U.S. corn, Brazilian sugarcane), while oil majors have historically resisted blending mandates. The EU's Renewable Energy Directive (RED) reflects geopolitical energy independence goals as much as emissions targets.

71

## Policymakers must balance:

- Domestic energy security vs. global environmental standards

- Farmer livelihoods vs. emissions metrics

- Lobbyist influence vs. empirical performance

> *"Policy certainty isn't about locking in rules—it's about making stakeholders confident they'll be heard when rules evolve."*
> *— Minister Karima Desai, Deputy Minister for Renewable Energy and Climate Affairs*

As such, policy isn't just technical—it's negotiated, reinterpreted and vulnerable to shifts in leadership, trade pressures or public sentiment.

Biofuel policy is shaped by a web of political and economic interests. Agricultural lobbies influence feedstock-specific incentives. Energy independence concerns dominate EU directives. Effective policy must balance climate, commerce and constituency demands.

## To improve effectiveness and stakeholder confidence, future policy should:

- Harmonize definitions and emission metrics across jurisdictions

- Link incentives to verifiable impact (e.g., net GHG reduction, water use efficiency)

- Support inclusive project development with community consultation protocols

## Challenges in Enforcement and Policy Durability

Even well-designed policies can falter without proper enforcement and long-term consistency, especially in fragmented or rapidly evolving regulatory landscapes.

Even the best-designed policies can fail without proper enforcement or long-term consistency. In India, for instance, the ambitious E20 ethanol target has faced infrastructure bottlenecks and regional feedstock shortages. In contrast, California's LCFS succeeded in driving innovation but required constant updates to credit eligibility rules, causing regulatory fatigue among some producers.

## Recommendations:

- Pair incentives with independent verification (e.g., blockchain + third-party audits)

- Include sunset clauses and review cycles in mandates

- Strengthen interagency coordination (transport, agriculture, energy, environment)

Policies often fail not due to poor design but because of enforcement gaps or inconsistent application. India's ethanol targets are delayed by feedstock and logistics issues. California's LCFS, though effective, requires constant updates and rule clarification. Strong governance and flexible design are essential.

## Concluding Insights

Future policy must reward innovation while minimizing harm, and align with both market and environmental needs. Effective frameworks treat regulation as a dynamic process, not static legislation.

### Visual Scorecard: Comparative Policy Effectiveness

The following radar chart illustrates the comparative strengths of different biofuel policy mechanisms across five key criteria: market impact, innovation support, scalability, policy stability and environmental integrity. It is designed to provide a visual summary of how each policy balances tradeoffs in practical deployment.

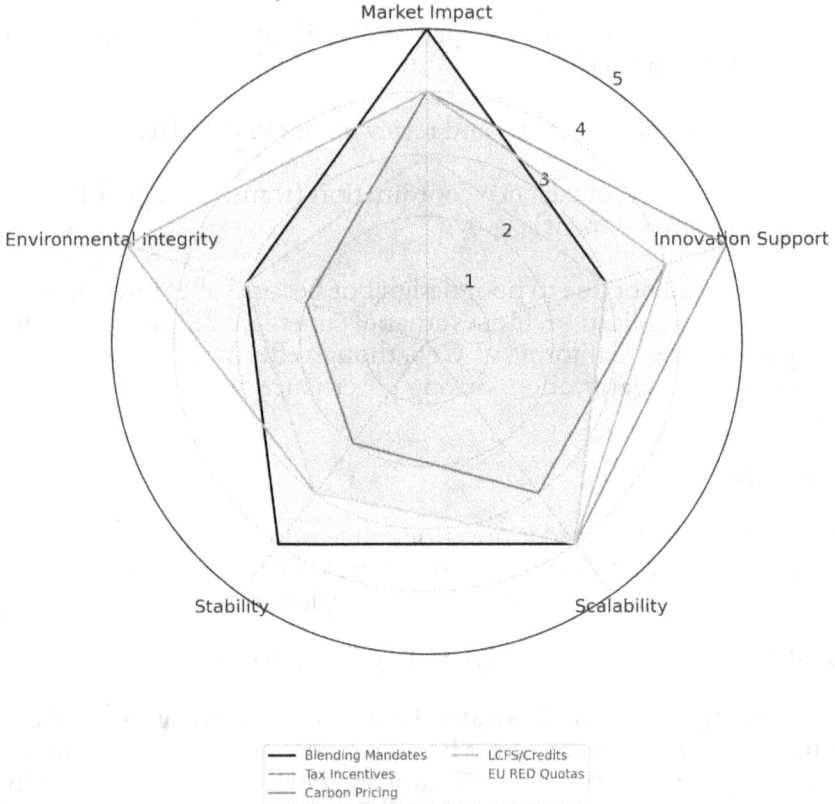

**Biofuel Policy Effectiveness Scorecard**

*Radar chart showing comparative performance of selected biofuel policy mechanisms. Higher values indicate stronger performance across each policy dimension.*

While these policy mechanisms have driven progress, their implementation has also produced a range of unintended environmental, social and economic outcomes.

Despite well-meaning objectives, many biofuel policies have led to unintended economic, environmental and social consequences.

**For example:**

- Corn ethanol mandates in the U.S. contributed to food price inflation during the 2008 crisis, raising concerns over "food vs. fuel" tradeoffs.

- Palm-based biodiesel subsidies in Southeast Asia created perverse incentives for deforestation, undermining the GHG benefits of the fuel.

- Volatile LCFS credit prices have caused market instability, making it difficult for smaller producers to plan long-term investments.

Such effects underscore the need for adaptive policy design that includes lifecycle emissions accounting, indirect land-use change (ILUC) modeling and mechanisms to reduce price and compliance volatility.

# Regulatory Changes and Risk Management

In the rapidly evolving biofuel sector, effective risk management and strategic planning for regulatory changes have become paramount. Organizations frequently underestimate the complexity and impact of regulatory shifts on their operations.

Policy volatility—whether through subsidy phaseouts, shifting sustainability thresholds or trade restrictions—can expose biofuel projects to significant risk.

A comprehensive risk management framework for regulatory changes should address both immediate compliance requirements and potential future developments. This approach necessitates organizations to maintain robust regulatory monitoring systems while developing adaptive strategies that can respond to evolving regulatory landscapes and maintain operational flexibility.

The following key elements form the foundation of effective **regulatory risk management**:

- **Regulatory Intelligence Systems:** Mechanisms for monitoring and analyzing policy developments

- **Impact Assessment Frameworks:** Tools for evaluating potential effects of regulatory changes

- **Stakeholder Engagement Strategies:** Programs for maintaining dialogue with regulatory bodies

- **Compliance Management Systems:** Integrated approaches to maintaining regulatory adherence

- **Change Management Protocols:** Procedures for implementing regulatory adaptations

- Build **multi-jurisdictional** compliance pathways

- **Diversification:** Markets and credit revenue streams

- **Maintain** regulatory intelligence systems

Successful regulatory risk management often depends more on organizational preparedness than reactive responses.

**Case Example**
*A U.S. cellulosic ethanol project facing fluctuating RIN values hedged its exposure by entering long-term offtake agreements indexed to both oil prices and LCFS credits. This dual-credit strategy stabilized cash flow while satisfying investors' ESG expectations.*

**Case Example**
*A project involving the transition of a large industrial facility to biofuel systems, a three-tier approach to managing regulatory risk was implemented:*
*\* immediate compliance*
*\* medium-term adaption and*
*\* long-term strategic positioning*
*This structured approach proved particularly valuable when new emissions standards were introduced midway through the project, as clear protocols for regulatory adaptation were established. This allowed the team to quickly assess the impact and implement necessary adjustments without significant disruption to operations.*

A crucial aspect of strategic planning involves building relationships with regulatory bodies and industry associations to gain valuable insights into potential policy developments and opportunities to contribute to the regulatory dialogue. However, it's essential to maintain appropriate boundaries and ensure all interactions comply with ethical guidelines and transparency requirements.

*"Local opposition often stems from not being heard—not from being against biofuels. Engagement has to be proactive, not reactive."*

— Thomas Leong, Former City Council Member

As we wrap up our discussion on regulatory frameworks in the biofuel sector, it becomes evident that navigating the complex policy landscape requires more than mere compliance—it demands strategic foresight and adaptable management systems. The complexities of navigating biofuel market regulations at various levels present both obstacles and benefits.

Throughout this chapter, we examined the different facets of regulatory compliance, including certifications, policies and risk. The importance of developing integrated approaches to managing these requirements has been highlighted, with practical examples from industrial applications reinforcing this point.

It is essential to keep in mind that effectively navigating regulatory frameworks necessitates ongoing learning and adaptation. Organizations need to stay alert to policy changes and be flexible enough to adjust to new requirements as they arise.

# Key Takeaways

- **Regulatory compliance is a strategic advantage**, not just a legal obligation. Organizations that proactively engage with evolving biofuel regulations and certification standards can build more resilient, competitive and sustainable operations.

- **Global certifications** such as ISCC and RSB ensure sustainability across the value chain but require integrated systems, digital documentation and ongoing verification to manage complexity and meet diverse market expectations.

- **Policy mechanisms and incentives** (e.g., blending mandates, tax breaks, carbon pricing) vary widely across regions, demanding adaptable strategies and active monitoring to align operations with changing market conditions.

- **Multi-jurisdictional regulatory risk management** must go beyond compliance tracking—successful organizations use forward-looking models, scenario planning, and stakeholder engagement to anticipate changes and minimize disruption.

- **Unified compliance frameworks**, such as cross-border compliance matrices, simplify multi-jurisdictional challenges, reduce redundancy and create a foundation for scalable, global biofuel operations.

Chapter 6

# Implementation Roadmap

*Integration Strategies for Industrial Transition*

This chapter is tailored for operations managers, systems engineers and technical project leads overseeing the transition to biofuels in industrial contexts. It provides actionable strategies, international case examples and infrastructure planning tools for real-world deployment.

The journey toward industrial biofuel adoption goes beyond technological readiness; it demands a carefully orchestrated implementation strategy that addresses every aspect of the transition process. A thorough plan that considers technicalities, team collaboration, stakeholder needs and operational stability is key to success. This understanding serves as the foundation for developing effective transition strategies that can turn theoretical possibilities into practical realities. Implementing biofuel solutions requires a carefully orchestrated approach that balances technical requirements, operational continuity and stakeholder engagement in the rapidly evolving landscape of industrial energy systems.

This chapter delves into the essential components of a successful implementation roadmap, exploring how organizations can navigate the complexities of biofuel adoption while maintaining operational excellence. It will examine proven methodologies for developing comprehensive transition plans, establishing effective monitoring systems and managing change across diverse stakeholder groups, focusing on practical, actionable strategies that can be adapted to various industrial contexts.

Through detailed analysis of implementation frameworks and real-world case examples, organizations can learn how to develop robust transition plans that address both technical and human factors. Key focus areas include examining critical success factors, such as timeline development, resource allocation and performance monitoring systems, with special attention given to strategies for managing resistance to change and building stakeholder support throughout the implementation process.

## Biofuel Transition Phases and Risk Zones

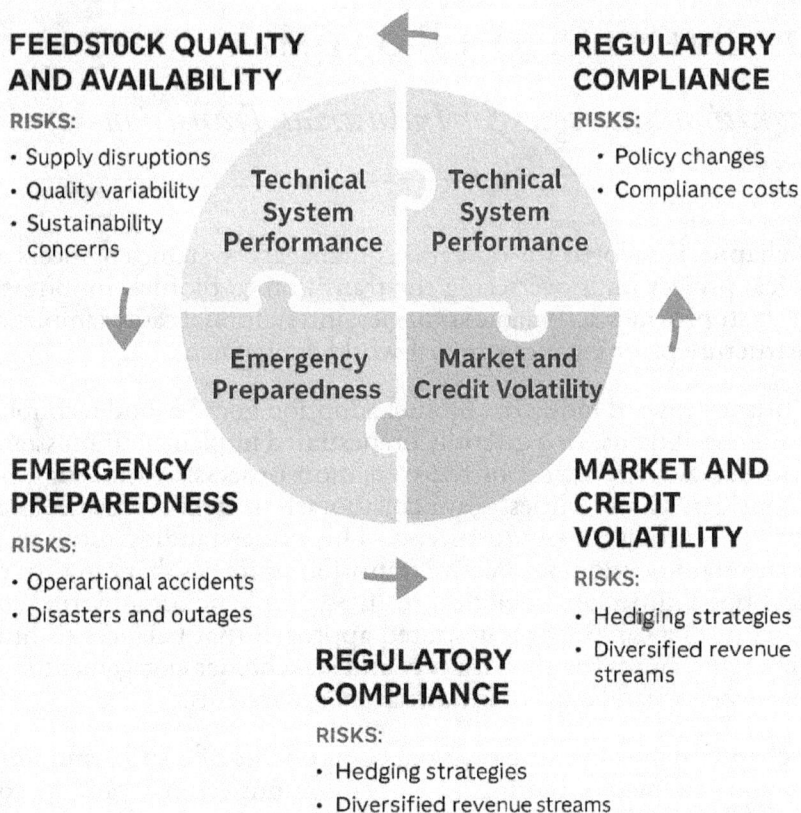

**FEEDSTOCK QUALITY AND AVAILABILITY**

RISKS:

- Supply disruptions
- Quality variability
- Sustainability concerns

Technical System Performance

**REGULATORY COMPLIANCE**

RISKS:

- Policy changes
- Compliance costs

Technical System Performance

Emergency Preparedness

Market and Credit Volatility

**EMERGENCY PREPAREDNESS**

RISKS:

- Operartional accidents
- Disasters and outages

**MARKET AND CREDIT VOLATILITY**

RISKS:

- Hedging strategies
- Diversified revenue streams

**REGULATORY COMPLIANCE**

RISKS:

- Hedging strategies
- Diversified revenue streams

The insights provided in this chapter are particularly relevant for organizations seeking to decarbonize their operations while maintaining competitive advantage, as understanding the intricate balance between technical implementation and change management will better equip readers to lead successful biofuel transitions in their own organizations. This roadmap explains how to handle the challenges and benefits of changing how industry uses energy.

*"Even minor disruptions in distribution can break trust with end-users," explains Luis Ortega, Procurement Director*

### Case Example: Transition to HVO in Remote Mining Operations

*The complexity of industrial transitions is exemplified by a recent case that arose during a major facility's shift to biofuel systems.*

*Facing carbon mandates and remote fuel delivery costs, a mining operator initiated a transition to Hydrotreated Vegetable Oil (HVO). Initial resistance from maintenance crews centered on fuel stability and engine longevity. The transition used a phased strategy, beginning with a tightly controlled pilot.*

## Pilot Phase

- **Scope:** Five Caterpillar 793F haul trucks

- **Infrastructure Modifications:** Only minor seal replacements required

- **Monitoring Tools:** Engine sensors, fuel flow meters, emissions analyzers

- **Results (after 90 days):**

  - $CO_2e$ emissions reduced by 82%

  - No significant change in engine wear

  - Fuel efficiency drop of ~2.5% offset by bulk pricing

## Engagement Approach:

- Daily feedback loops and dashboard displays to reduce uncertainty

- Cross-functional training by HVO supplier and OEM representatives

- Maintenance teams consulted early and continuously

## Scale-Up Phase:

- 80% fleet conversion in 9 months

- On-site blending and storage installation with smart monitoring

- Emissions disclosures used to strengthen ESG reporting

**Impact:**

- 64% sitewide emissions reduction year-over-year

- Zero unplanned downtime throughout the transition

**Example:**

| Parameter | Baseline (USLSD) | HVO (Biofuel) | Delta |
|---|---|---|---|
| Fuel Cost ($/gal) | $3.35 | $3.75 | +$0.40 |
| Fuel Efficiency (MPG equiv.) | 5.1 | 4.97 | -2.5% |
| Maintenance Cost/Month | $48,000 | $44,000 | -$4,000 |
| Carbon Credit Value/Mo | $0 | $18,500 (est.) | +$18,500 |
| Net Monthly Impact | – | – | +~$14,500 gain |

*\* Data assumptions drawn from California LCFS, 2023 fuel pricing benchmarks, and Caterpillar maintenance data ranges.*

## Technology-Specific Implementation Economics

Cost-benefit snapshot comparing different industrial biofuel transition types.

| Use Case | Biofuel | CapEx (est.) | Opex Delta | Payback | Notes |
|---|---|---|---|---|---|
| Mining trucks (diesel swap) | HVO | Low | +5–10% | 2–4 yrs | High ESG benefit; minor retrofitting required |
| Boiler steam (heating fuel) | Bio-oil | Med | Neutral | 3–5 yrs | Good fit for food and pulp industry |
| Aviation (SAF for turbofans) | HEFA | Very High | +40–60% | 5–8 yrs | Needs offtake + policy support |
| Fleet buses (urban transit) | RD/Biodiesel | Low | Slight savings | 2–3 yrs | Low tech risk; proven in cold climates |

*\* Reference: IEA Renewables 2023; DOE BETO MYPP 2023*

## Phased Implementation Planning: From Pilot to Full-Scale

Implementing biofuel systems in industrial settings requires a carefully structured, phased approach to minimize operational disruptions and maximize adoption success. The development of comprehensive transition timelines and clear milestones is essential in laying the foundation for this process. These tools enable organizations to track progress and make necessary adjustments to their strategies.

The phased implementation typically begins with a detailed assessment phase, where current operations are analyzed and baseline performance metrics are established. This initial phase should include comprehensive energy audits, equipment compatibility assessments and stakeholder mapping to identify potential challenges and opportunities. The data collected during this phase informs the development of realistic timelines and achievable milestones.

Before initiating any transition, organizations must conduct a multi-layered readiness assessment.

**Key Diagnostic Domains:**

- **Feedstock supply chain audit:** Availability, seasonality, pricing risk

- **Facility infrastructure mapping:** Storage compatibility, distribution logistics

- **Technology interface check:** Engine/system compatibility with biodiesel, ethanol, SAF

- **Regulatory gap analysis:** Environmental permitting, safety standards

- **Staff readiness:** Skills audit, training requirements

Incorporating key elements into the implementation timeline is crucial for the success of the biofuel system integration.

**Typical Timeline Milestones:**

- **Assessment Phase (1-3 months):** Audits, feasibility studies, baseline metrics

- **Pilot Phase Implementation (3-6 months):** Focused testing with robust data collection

- **Expansion Phase (6-18 months):** Iterative rollout based on pilot outcomes

- **Full-Scale Integration:** Supply chain contracts, integration into ESG reporting metrics

## Key Inclusions:

- Training and development program rollouts and certifications

- Real-time performance dashboards (fuel consumption, emissions)

- Infrastructure Modification Schedule

- Supply Chain Integration Milestones

The Pilot Phase is a crucial learning opportunity that allows organizations to test assumptions and refine processes before broader implementation. Focusing on a single production line or facility area during this phase provides valuable insights while minimizing risk. It is important to have clearly defined success metrics, including factors such as fuel efficiency, equipment performance and operational reliability, which should be monitored closely.

Once the Pilot Phase demonstrates success, the implementation can be expanded in carefully planned stages. Each expansion phase should build upon lessons learned from previous stages, with clear transition points and success criteria. This approach allows organizations to maintain operational stability while systematically expanding their biofuel adoption.

### Case Example: Canadian Freight Rail
*A Class 1 railway tested B20 biodiesel in six locomotives over a harsh winter season. With engine monitoring telemetry and close coordination with OEMs, the pilot reported no cold-start failures. This led to a broader rollout across regional yards within 18 months.*

Stakeholder engagement is a critical component of the implementation timeline, with key stakeholders such as operations teams, maintenance staff and management being contacted regularly. These interactions help build support for the transition and provide valuable feedback that can be incorporated into the implementation process.

Milestone development should follow the SMART criteria - Specific, Measurable, Achievable, Relevant and Time-bound. Effective milestones to consider include:

- Limited scope (e.g., 10% of the fleet)

- Real-time monitoring and benchmarking against diesel baselines

- 30% Fleet / Equipment Modifications by month 6

- 100% Staff Training Certification by month 6

- 10% improvement in fuel-related downtime by month 9

- Achievement of Target Fuel Blend Ratios

- Supply Chain Integration Benchmarks

- Environmental Performance Targets

The implementation timeline should also incorporate contingency planning and flexibility to address unexpected challenges, while the plan needs clear milestones to track progress and make changes as needed. Regular review allows for evaluating progress against established milestones and refining future implementation phases, with performance monitoring systems being vital for tracking progress against established milestones.

## Risk Matrix for Biofuel Transitions (Industrial Sector)

| Risk | Likelihood | Impact | Mitigation Strategy |
|---|---|---|---|
| Feedstock supply disruption | Medium | High | Secure multi-year contracts; local diversification |
| Tech underperformance (pilot) | Medium | High | Staged deployment; vendor guarantees |
| Capex overrun (storage/blending) | High | Medium | Detailed site audits; contingency budgeting |
| Regulatory delays (credit flow) | Low–Medium | High | Buffer periods, multi-jurisdictional deployment |

These systems should capture both technical and operational metrics, providing real-time feedback on the implementation's success. Key performance indicators (KPIs) should be established early in the planning process and monitored throughout the implementation phases.

Resource allocation is another critical aspect of implementation planning, and the timeline should clearly outline when specific resources, including personnel, equipment and financial investments, will be required. Clearly defining resource allocation ensures that necessary resources are available when needed and prevents delays in implementation progress.

To effectively capture lessons learned and best practices during the implementation process, it is essential to establish documentation and knowledge management systems. This information plays a crucial role in supporting future implementation phases and aiding other facilities within the organization in their transition efforts.

As the implementation progresses, it is essential to conduct regular assessments to assess the effectiveness of the transition strategy. These assessments should cover technical performance metrics as well as organizational factors like staff engagement and operational efficiency. The insights gained from these assessments can be utilized to fine-tune the implementation approach for subsequent phases.

## Stakeholder Management and Change Leadership in Biofuel Integration

To successfully integrate biofuels into industry, effective stakeholder management along with strong leadership are essential. Since transition to biofuel systems affects numerous stakeholders across the organization, ranging from operational staff to senior management, it is important to recognize that each group has unique concerns and perspectives that must be carefully addressed to ensure project success.

To achieve this, a comprehensive stakeholder management strategy should begin with thorough stakeholder mapping and analysis. Typically, key stakeholder groups include:

- Operations and maintenance teams

- Senior management and board members

- Supply chain partners

- Regulatory compliance officers

- Local community representatives

- Environmental health and safety teams

*"You can't build a national biofuel strategy in a vacuum. You need farmers, freight carriers, environmental groups—and a lot of pa_ tience."*

*— Minister Karima Desai, Deputy Minister for Renewable Energy and Climate Affairs*

Understanding each stakeholder group's unique concerns and motivations enables the development of targeted engagement strategies. For example, operations teams often prioritize reliability and performance, while senior management may focus on cost implications and return on investment. To address these diverse perspectives, clear communication channels and tailored messaging that align with each group's priorities are essential.

Change leadership in biofuel integration requires a structured approach that combines technical expertise with strong people management skills. Successful transitions often involve a dedicated change management team working alongside technical implementation teams to ensure that both the technical and human aspects of the transition receive adequate attention and resources. Effective change leadership strategies should incorporate several key elements:

- Clear vision communication

- Transparent decision-making processes

- Regular progress updates

- Feedback mechanisms

- Recognition of early adopters

- Support systems for affected staff

Transitioning a large industrial facility to biofuel systems highlights the importance of proactive stakeholder engagement. The initial resistance from maintenance teams, who were worried about equipment reliability and maintenance procedures, was effectively managed through a combination of comprehensive training programs and hands-on demonstration projects. This approach not only enhanced technical competence but also fostered ownership and commitment to the transition process.

It is crucial to understand that stakeholder engagement should be seen as an ongoing process rather than a one-time effort. Regular communication touchpoints, feedback sessions and progress updates play a vital role in maintaining momentum and addressing concerns as they arise. This continuous engagement approach allows organizations to identify and tackle potential issues before they evolve into significant obstacles to implementation.

When integrating biofuels, leadership changes should take into account company culture and current work methods.

**Successful transitions often entail:**

- Cultural assessment and alignment

- Skills gap analysis and training programs

- Performance management adjustments

- Recognition and reward systems

- Knowledge transfer mechanisms

The development of change champions within different stakeholder groups can significantly enhance the effectiveness of transition efforts by serving as bridges between the implementation team and their respective stakeholder groups. They provide valuable insights and help build support for the transition at all organizational levels.

Measuring the effectiveness of stakeholder management and change leadership efforts requires both quantitative and qualitative metrics, such as stakeholder satisfaction scores, training completion rates, and the number and nature of implementation issues raised and resolved. Regularly assessing these metrics enables organizations to refine their approach and maintain momentum throughout the transition process.

**Case Example: Sustainable Aviation Fuel (SAF) Integration in Nordic Regional Aviation**
*A Nordic regional airline introduced a 30% SAF blend on domestic routes. The phased rollout began with short-haul flights, requiring minimal aircraft modifications.*
**Strategies Used:**
*\* Partnered with a SAF supplier for lifecycle emissions data*
*\* Communicated carbon savings directly on passenger tickets*
*\* Trained mechanics on storage and temperature considerations*
**Results:**
*\* 45% $CO_2e$ reduction per flight segment*
*\* On-time performance maintained*
*This example underscores the value of cross-sector engagement—operations, marketing, and compliance teams collaborated to build trust and awareness internally and externally.*

Successful change leadership also involves preparing the organization for ongoing evolution in biofuel technologies and operational practices. This forward-looking approach helps build organizational resilience and adaptability, essential qualities in the rapidly evolving landscape of sustainable energy solutions.

Stakeholder management and change leadership in biofuel integration represent ongoing commitments rather than finite projects, as robust stakeholder engagement ensures successful and sustainable organizational change.

## Public Awareness and Consumer Education: The Missing Link in Biofuel Adoption

While industrial transition strategies focus on technical implementation and stakeholder coordination, public acceptance and awareness are often overlooked.

Why Public Perception Matters:

Consumer skepticism can stall adoption, even when the infrastructure is ready.

Strategies for Effective Outreach:

- Simplify the science with visuals and analogies

- Localize the message using real-world examples

- Leverage trusted messengers like educators and transit authorities

- Show the co-benefits: air quality, job creation, energy security

Public Education and Consumer Awareness

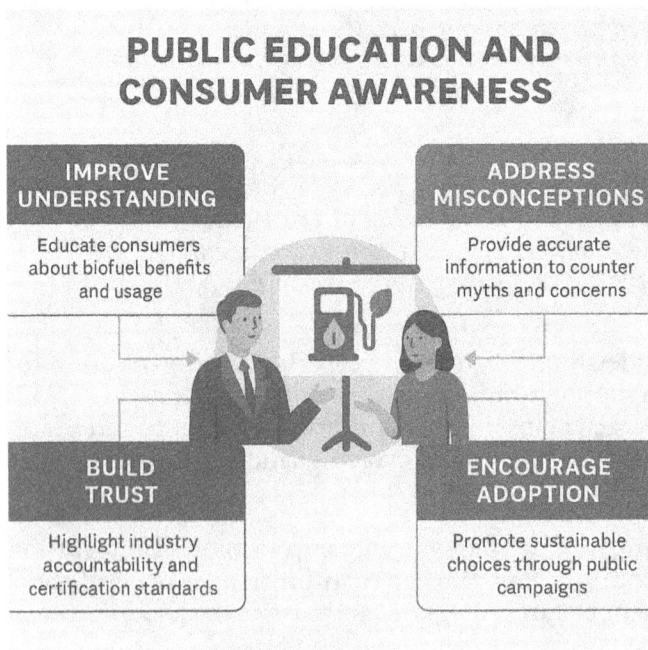

**PUBLIC EDUCATION AND CONSUMER AWARENESS**

**IMPROVE UNDERSTANDING**
Educate consumers about biofuel benefits and usage

**ADDRESS MISCONCEPTIONS**
Provide accurate information to counter myths and concerns

**BUILD TRUST**
Highlight industry accountability and certification standards

**ENCOURAGE ADOPTION**
Promote sustainable choices through public campaigns

**Case Example: Brazil's Biofuel Literacy Campaign**
*- National effort with school modules, public TV and interactive fuel station kiosks*
*- Ethanol acceptance remained high as blends increased.*

Opportunities for Industrial Stakeholders:

- Include public education in ESG reports.

- Involve consumers in pilot projects with transparent labeling.

- Co-brand public transit programs with biofuel performance data.

# Performance Monitoring and Continuous Improvement Frameworks

Effective performance monitoring and continuous improvement frameworks form the backbone of successful biofuel implementation strategies in industrial settings. By systematically tracking progress, these systems not only boost biofuel technology but also drive ongoing improvements in efficiency and sustainability.

To achieve this, effective monitoring requires careful selection of key performance indicators (KPI) across all operational areas.

**Key metrics typically include:**

- Fuel efficiency and consumption patterns

- Real-time process parameter tracking and analysis

- Equipment condition monitoring and predictive maintenance indicators

- Product quality monitoring and trend analysis

- Emissions reduction measurements

- Cost performance indicators and optimization metrics

- Supply chain reliability indicators

- Quality control benchmarks

To establish a robust monitoring framework, it is essential to combine real-time data collection with long-term trend analysis. This approach allows organizations to address immediate operational challenges promptly while maintaining a strategic focus on improvement goals.

**Monitoring Infrastructure may include:**

- Telematics for mobile fleets

- IoT-based diagnostics for combustion efficiency

- Fuel blending sensors

- Emissions tracking via AI-enhanced data streams

- Digital twins for scenario modeling

By leveraging advanced sensors and data analytics, modern monitoring systems can provide accurate and timely information.

**Case Example**
*A large industrial facility facing issues with inconsistent biofuel performance across various equipment types successfully implemented a comprehensive monitoring framework. Through tracking key parameters across all operational units, the system identified previously unnoticed patterns in equipment performance. This insight enabled the facility to make targeted optimizations that significantly enhanced overall efficiency.*

To sustain progress in biofuel operations, a structured approach that capitalizes on performance monitoring insights is crucial. The Plan-Do-Check-Act (PDCA) cycle serves as an effective framework for implementing continuous improvements.

- **Plan:** Identify opportunities for optimization based on monitoring data

- **Do:** Implement targeted improvements in controlled phases

- **Check:** Measure and analyze the results of improvements

- **Act:** Standardize successful changes and identify next improvement areas

Data integration plays a crucial role in effective performance monitoring. Modern systems should merge information from multiple sources, including:

- Equipment sensors and control systems

- Laboratory analysis results

- Maintenance records

- Supply chain data

- Environmental monitoring systems

Establishing clear performance baselines is essential, as it enables organizations to measure accurately the impact of improvement initiatives. To remain effective, these baselines should be regularly reviewed and updated to account for changing operational conditions and technological advancements. Furthermore, regular performance reviews, conducted at predetermined intervals, help ensure that improvement efforts remain aligned with organizational objectives.

In addition to baseline assessments, successful continuous improvement frameworks must also incorporate feedback mechanisms that allow for a rapid response to operational challenges. To achieve this, key components for these frameworks include establishing clear protocols for:

- Performance deviation alerts

- Root cause analysis procedures

- Corrective action implementation

- Results verification methods

- Knowledge sharing systems

Despite advancements in technology, the human element remains crucial to performance monitoring and improvement efforts. By fostering this understanding, organizations can build a culture of continuous improvement where employees actively participate in identifying and implementing optimization opportunities.

Moreover, documenting project improvements creates valuable knowledge for future projects. To be truly effective, this documentation should include detailed records of:

- Improvement project objectives

- Implementation methodologies

- Results and outcomes

- Lessons learned

- Best practices identified

To maintain accuracy and efficiency, monitoring systems should be regularly reviewed and refined to ensure their continued relevance and effectiveness. Specifically, regularly checking monitoring parameters, data collection methods and analysis techniques helps identify enhancement opportunities while also ensuring optimal performance. As technology continues to evolve, monitoring systems should be updated to incorporate new capabilities that provide deeper insights into operational performance.

Furthermore, integrating predictive analytics and machine learning capabilities can significantly enhance the effectiveness of performance monitoring systems.

**By leveraging these advanced tools, organizations can:**

- Identify potential issues before they impact operations

- Optimize maintenance scheduling

- Predict equipment performance trends

- Enhance resource allocation efficiency

- Improve decision-making processes

Success in biofuel implementation ultimately depends on continuously monitoring, analyzing and improving operational performance. To succeed with biofuels long-term, organizations need strong monitoring and ongoing improvement. Implementing biofuel systems in industrial settings represents a complex but achievable transition that demands careful planning, stakeholder engagement and systematic execution. Beyond technical skills, successful implementation needs a broader strategy addressing organizational factors, operational continuity and change management, as discussed in this chapter.

**Continuous Improvement in Practice: Bus Fleet**
*A transit agency piloted renewable diesel (RD) on 15 buses. Real-time fuel flow and emissions monitoring identified cold-weather viscosity issues. Based on findings, the agency adjusted storage protocols and optimized blending ratios seasonally.*

# The Complex Landscape of Biofuel Blending

Blending biofuels with conventional fossil fuels is one of the most practical ways to introduce renewable energy into existing energy systems. It allows gradual decarbonization without overhauling infrastructure or engines, making it a pivotal strategy in sectors like transportation, aviation and heavy industry.

## Blending Ratios and Applications

Common Blends:

- E10, E15, E85: Ethanol-gasoline blends used primarily in light-duty vehicles

- B5, B20, B100: Biodiesel blends commonly used in trucking, agriculture and municipal fleets

- SAF Blends (up to 50%): Used in commercial and military aviation, requiring strict certification

Sectoral Applications:

- Road Transport: Most vehicles can use E10 or B5 without modification

- Aviation: SAF must meet ASTM D7566; blending capped due to certification limits

- Marine: Blending trials are ongoing, with B20 often used in inland shipping

Note: Blending mandates vary by region, from Brazil's B12 biodiesel standard to the EU's RED II requirements.

## Technical Challenges

Fuel Compatibility:

- Engine and fuel system components may degrade with higher biofuel content

- Water absorption and microbial growth in biodiesel blends (esp. B20-B100)

Cold Flow Properties:

- Higher biodiesel blends can gel in cold climates without additives

- Ethanol has a lower energy density, impacting range

Emissions Control:

- Blends can affect NOx, PM and CO2 profiles differently

- Compatibility with emission after-treatment systems (e.g., DPFs, SCRs)

## Regulatory Frameworks and Blending Mandates

International Examples:

- Brazil: Mandatory B12 biodiesel; E27 ethanol-gasoline blend

- United States: RFS requires Renewable Identification Numbers (RINs) tracking

- European Union: RED II targets 14% renewables in transport by 2030

- India: National Bio-Energy Policy targets 20% ethanol by 2025

Compliance Requirements:

- Accurate tracking of blend ratios

- Certification and testing per ASTM, EN, or ISO standards

- Lifecycle emissions reporting (e.g., GREET, GHGenius models)

## Infrastructure and Quality Control

Key Infrastructure Needs:

- Dual-fuel pumps and storage tanks with corrosion-resistant linings

- Temperature-controlled blending facilities

- Dedicated rail and truck logistics for high-blend transport

Quality Assurance Protocols:

- Routine blend verification testing

- Water content and cloud point analysis (critical for biodiesel)

- Monitoring for phase separation in ethanol blends

Blending Best Practices

- Use in-line blending systems for real-time ratio adjustment

- Implement seasonal blend switching (e.g., B5 in winter, B20 in summer)

- Monitor microbial growth in storage tanks quarterly

- Co-locate labs with blending terminals for faster quality assurance

## Market Dynamics and Economic Considerations

Pricing Volatility:

- Biofuel prices linked to commodity markets (e.g., corn, soy, palm oil)

- Credit systems like RINs (U.S.) or LCFS (California) offset blend costs

Supply Chain Integration:

- Refinery co-processing vs. terminal blending

- Regional variability in blend stock availability

### Case Example: California LCFS
*Blenders receive credits for low-carbon-intensity fuels. Strategic blending (e.g., E85 in flex-fuel vehicles) yields strong economic returns due to the stacking of RIN and LCFS credits.*

### Biofuel Blending Ecosystem

**BIOFUEL BLENDING ECOSYSTEM**

| Blending Ratios | Blending Infrastructurre | Regulatory Frameworks |
|---|---|---|
| E10 | | • Blending Mandates |
| B20 | | • Fuel Quality Requirements |
| E85 | Storage Dispensing Refining / Mixing | • Harmonization of Standards |
| B100 | | |

**Quality Control Protocols**
- Fuel Properties Testing
- Contamination Prevention
- Certification Standards
- Regular Monitoring

**Regulatory Frameworks**
Transportation
Industrial Fuels
- Blending Mandates
- Fuel Quality Requirements

Biofuel blending is more than a technical workaround; it's a transitional pathway that links current infrastructure with a low-carbon future. Success hinges on harmonizing policy, technology and quality control. Addressing blending complexities today will accelerate adoption and unlock the full potential of biofuels tomorrow.

## Key Takeaways

While the pathway to industrial biofuel adoption is promising, it is not without hurdles. Biofuels offer compelling sustainability and operational benefits, but successful transitions require a clear-eyed understanding of the risks and trade-offs.

- **Start small, scale smart**: Pilot projects reduce reputational and financial risk but must be grounded in thorough diagnostics.

- **Invest in people**: Technical readiness is only part of the equation—successful transitions hinge on workforce engagement and reskilling.

- **Monitor continuously**: Feedback loops must not only detect problems but also support course corrections and long-term improvement.

- **Think globally, deploy locally**: While international case examples from Canada and Northern Europe offer valuable insights, every industrial context presents unique constraints—be it infrastructure, policy or climate.

---

■ *Risks to Watch: Common Pitfalls in Biofuel Implementation*
* *Inconsistency: Seasonal fluctuations and local sourcing constraints can disrupt supply and alter fuel performance.*
* *Underestimated Infrastructure Costs: Storage, filtration, and handling systems often need more extensive upgrades than initially planned.*
* *Overlooking Training Gaps: Even minor procedural errors in handling biofuels (e.g., moisture exposure) can compromise performance or safety.*
* *Policy Shifts: Changes in subsidy programs or credit structures (e.g., LCFS, RINs) can impact ROI projections mid-project.*
* *Technology Overreach: Pushing unproven blends (e.g., B100) without sufficient data or cold-weather testing can lead to reliability issues.*

---

As industrial operations look toward biofuels as a pillar of decarbonization, they must do so with both ambition and caution. Implementing with resilience means being prepared for variability in feedstock quality, fluctuations in policy incentives and emerging technical limitations. A successful transition strategy acknowledges these realities while still pushing toward scalable, low-carbon solutions.

- **Phased implementation**—starting with pilot programs and building up in controlled stages—is the most effective strategy for minimizing risk and ensuring operational continuity during biofuel adoption.

- **Stakeholder engagement and change leadership** are essential. Successful transitions require clear communication, role-based training, and proactive involvement of all key teams—from operations to executives.

- **Real-time performance monitoring systems** (including KPIs for emissions, fuel efficiency, and equipment reliability) enable data-driven decision-making and continuous improvement throughout the transition.

- **Adaptive planning and SMART milestones** (e.g., fleet conversion rates, staff training benchmarks, emissions reduction targets) ensure projects stay on track while remaining flexible to unexpected challenges.

- **Case studies** from sectors like mining, aviation, and public transport show that integrating biofuels is not just technically feasible—it can also deliver rapid, measurable gains in emissions reduction and operational efficiency when properly executed.

Chapter 7

# Risk Mitigation

## *Managing Technical and Operational Challenges*

Managing risk is crucial for the success of biofuel projects. The intersection of new technologies, evolving operational requirements and stringent safety standards creates a unique set of challenges that demand systematic risk identification and mitigation approaches. This fundamental reality underscores the necessity of implementing robust risk management frameworks that can adapt to dynamic biofuel operations. To address supply chain vulnerabilities and process inefficiencies, organizations need strategies that manage technical and operational risks while maintaining efficient production.

Variations in feedstock quality can cascade into significant operational disruptions if proper risk mitigation strategies are not in place, extending beyond simple quality control measures to encompass the entire value chain from initial sourcing to final application. The interconnected nature of modern biofuel operations means that risks in one area can quickly impact the entire system.

### Feedstock Quality Variation vs. Equipment Downtime Chart

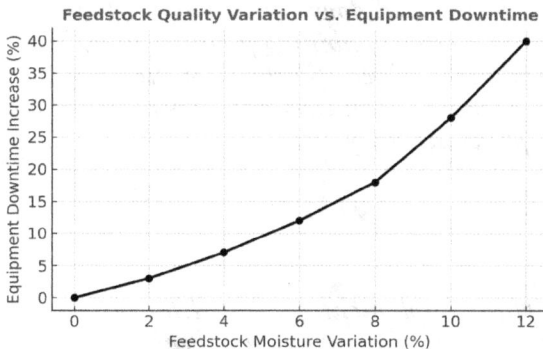

*Feedstock moisture variability impacts on equipment performance and reliability*

## Feedstock Risk and Quality Control

Variations in feedstock quality can cascade into significant operational disruptions. To manage this:

- Implement real-time feedstock quality testing and supplier audits

- Establish multi-sourcing strategies and regional buffer stocks

- Use moisture control, contamination tracking and nutrient balance metrics

## Risk Matrix: Feedstock Disruption

| Risk Factor | Likelihood | Impact | Mitigation Strategy |
|---|---|---|---|
| Crop yield shortfall | Medium | High | Weather modeling and sourcing diversity |
| Transport delays | Medium | Medium | Contract redundancy with logistics firms |
| Seasonal degradation | High | Medium | Regional storage and blending flexibility |

### Case Example
*A recent case highlights the critical need for comprehensive risk management in biofuel operations. The application of standard quality control procedures was essential to pinpoint the root cause and avoid increased downtime. A multi-layered risk management approach—combining feedstock analysis and real-time monitoring—proved effective in mitigating operational risks. Early warning indicators and response protocols helped maintain process stability.*

### Case Example
*A Brazilian ethanol plant implemented a satellite-linked feedstock monitoring system to track sugarcane health and harvesting patterns, reducing poor-quality feedstock deliveries by 38% and lowering fermentation failure rates.*

The intricacies of modern biofuel systems necessitate a systematic approach to risk assessment and mitigation. In addition to technological considerations, organizations must also factor in regulatory compliance and environmental impact when addressing system failures. This chapter delves into the essential aspects of managing risks in biofuel production, providing insights on how to identify, evaluate and mitigate potential risks effectively. Regulatory risks are covered in chapter 5.

Through the examination of real-world examples and proven methodologies, we will explore the essential elements of risk mitigation strategies. These strategies can assist organizations in maintaining operational reliability when adopting new biofuel technologies. Our focus will be on practical and implementable solutions that tackle both current challenges and emerging risks within the dynamic biofuel sector.

This chapter will highlight key risk management aspects, including quality control and emergency response. Understanding these elements is crucial for organizations looking to establish resilient biofuel operations capable of overcoming operational challenges and maintaining consistent performance standards.

## Technical Risk Assessment and Quality Control Systems in Biofuel Operations

A robust technical risk assessment framework, paired with quality control systems, forms the backbone of resilient biofuel operations.

**Key components include:**

- Feedstock quality verification protocols and testing procedures

- Process parameter monitoring and control systems

- Equipment performance tracking and preventive maintenance schedules

- Product quality assurance and certification processes

- Environmental impact monitoring and compliance verification

- Safety systems and emergency response protocols

> *"It's not that I oppose innovation—but let's be honest: biofuels are a lot more viable when you already have the capital, land, and infrastruc_ ture."*
>
> — Dario Fernández, Agronomist and Land Rights Advocate, Argentina

Implementing the above components requires careful consideration of both technical capabilities and operational constraints. For instance, feedstock quality verification must balance the need for thorough testing against production timeline requirements. Similarly, process monitoring systems as described in chapter 6 must provide comprehensive oversight while remaining practical for operators to manage effectively.

Quality control in biofuel operations extends beyond traditional product testing to encompass the entire production chain. Key aspects of this approach include incoming feedstock analysis, in-process testing and final product verification. Modern quality control systems increasingly incorporate real-time monitoring capabilities, allowing immediate detection and response to potential quality issues before they impact production outputs.

Risk assessment matrices are valuable tools for evaluating and prioritizing technical risks in biofuel operations. These matrices typically take into account both the likelihood of specific technical failures and their potential impact on operations, safety and environmental compliance. By systematically evaluating these factors, organizations can allocate resources more effectively to address the most critical risks first.

Another crucial element of technical risk management is the development of comprehensive standard operating procedures (SOPs). These procedures should encompass routine operations, quality control protocols and emergency response measures. Regularly reviewing and updating SOPs ensures they remain relevant as technologies and operational requirements evolve.

Competency development and training are vital for mitigating technical risks. To operate biofuel systems effectively, one must have a thorough understanding of the mechanics, chemistry and quality control involved. Regular training programs and practical experience help in building the expertise necessary for maintaining consistent operational standards.

Documentation and record-keeping systems track vital information, maintenance activities and test results, while also storing historical data for analysis and regulatory compliance. The use of digital platforms significantly improves the maintenance of comprehensive records, as well as enhances data analysis and reporting capabilities. By incorporating technical risk assessments and quality control measures, organizations can strengthen their biofuel operations, ensure consistent product quality and effectively manage operational risks. The key lies in implementing systematic approaches that integrate stringent technical standards with practical operational considerations.

## Biofuel Risk Matrix

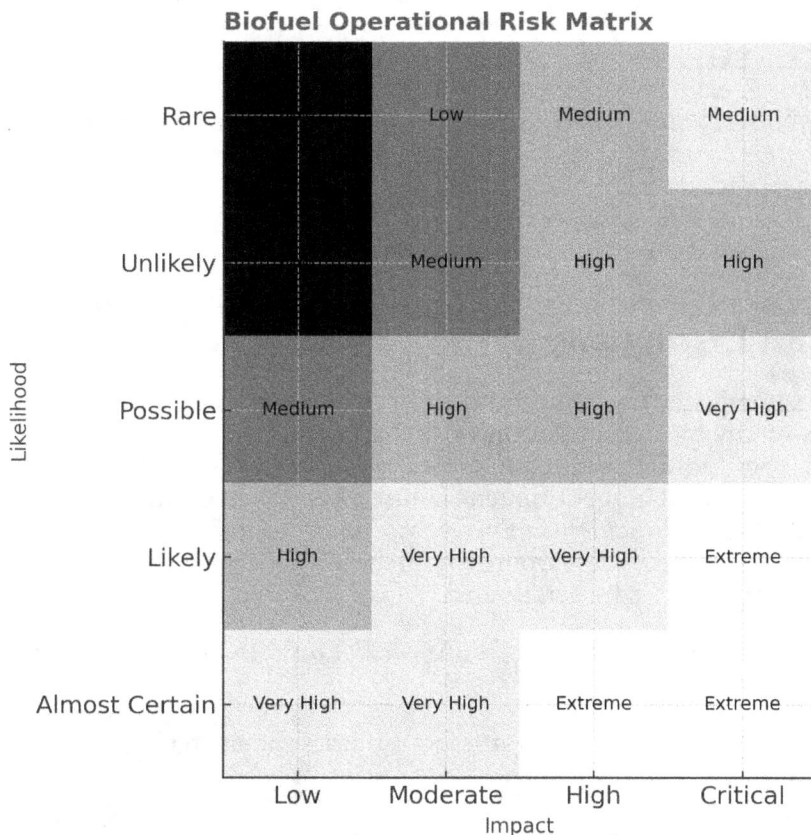

**Biofuel Operational Risk Matrix**

|  | Low | Moderate | High | Critical |
|---|---|---|---|---|
| **Rare** |  | Low | Medium | Medium |
| **Unlikely** |  | Medium | High | High |
| **Possible** | Medium | High | High | Very High |
| **Likely** | High | Very High | Very High | Extreme |
| **Almost Certain** | Very High | Very High | Extreme | Extreme |

Likelihood (vertical axis) / Impact (horizontal axis)

*\* Operational risk matrix illustrating relative risk levels in biofuel operations*

## Technology Performance and Process Reliability

Operational reliability depends on maintaining process stability across multiple units.

**Specific mitigation strategies include:**

- Deploy fault detection algorithms to preempt failures

- Maintain redundancy in key equipment (e.g., dual pumps, parallel reactors)

- Schedule predictive maintenance based on usage and wear data

**Case Example**
*A U.S. biodiesel plant reduced unplanned shutdowns by 22% after integrating infrared thermal monitoring and vibration sensors on centrifuge units, identifying early failure modes in bearings and seals.*

# Emergency Response and Contingency Planning for Operational Disruptions

Effective emergency response and contingency planning are critical components of any biofuel operation's risk management strategy. Given the complexity of biofuel production, operational disruptions can arise from various sources, such as equipment failures, supply chain interruptions and environmental incidents. Therefore, the key to maintaining operational continuity lies in developing comprehensive response protocols that can be rapidly deployed when needed.

**A robust emergency response framework should incorporate several essential elements:**

- Clearly defined roles and responsibilities for emergency response teams

- Step-by-step procedures for different types of operational disruptions

- Communication protocols for internal and external stakeholders

- Backup systems and redundancy measures for critical operations

- Regular training and simulation exercises

- Documentation and reporting procedures

One of the most valuable lessons is that effective contingency planning requires regularly reviewing and updating emergency procedures to ensure they evolve alongside technological advancements and changing operational requirements. For instance, emergency protocols must be updated when implementing new biofuel technology to account for specific hazards.

## Emergency Preparedness and Response

Facilities must prepare for acute events (e.g., chemical leaks, supply chain collapse, fire) through:

- Emergency drills with first responders

- Backup power systems and redundant communication lines

- Tiered response protocols with escalation triggers

### Case Example
*During the 2023 Gulf Coast freeze, a refinery maintained output using a mobile heating fleet and modular blending equipment staged onsite prior to winter.*

Another crucial aspect of contingency planning is developing backup supply chains by identifying alternative feedstock sources and maintaining relationships with multiple suppliers to ensure operational continuity during supply disruptions. This approach requires a careful balance between cost considerations and risk mitigation.

Regular emergency response drills and simulations are vital in maintaining readiness for potential disruptions, involving all relevant personnel and simulating various scenarios, from minor technical issues to major system failures. Through these drills, teams can identify potential gaps in response procedures and develop more effective solutions.

One of the most challenging aspects of emergency response is maintaining clear communication channels during crisis situations. Establishing predetermined communication protocols and ensuring all team members understand their roles can significantly improve response effectiveness.

Key measures include maintaining updated contact lists and establishing backup communication systems.

Integrating digital monitoring systems with emergency response protocols has revolutionized how organizations handle operational disruptions. These systems can provide early warning indicators and automated alerts, allowing for a more rapid response to potential issues. However, it's crucial to maintain manual backup procedures in case of technology failures.

Contingency planning must also address environmental and safety considerations. Key components include procedures for containing potential spills, managing emissions during system failures and ensuring worker safety during emergency responses. Environmental monitoring systems should be integrated with emergency response protocols to ensure compliance with regulatory requirements, even during disruptions.

Documentation and post-incident analysis are crucial components of effective emergency response systems. Organizations should keep detailed records of all incidents and responses, leveraging this information to enhance future contingency plans. This continuous improvement strategy contributes to the development of more resilient operations over time.

Financial planning for operational disruptions is another critical aspect of contingency planning. Organizations must maintain sufficient insurance coverage and emergency funds to address potential disruptions, aiming to minimize the impact on long-term operations. This financial preparedness also plays a vital role in upholding stakeholder confidence during challenging periods.

By implementing these comprehensive emergency response and contingency planning measures, organizations can fortify their biofuel operations to withstand various operational challenges. The key to success lies in establishing systematic, well-documented approaches that can be executed effectively under pressure, all while prioritizing safety and environmental compliance.

## Preventive Maintenance and Operational Resilience

Effective preventive maintenance strategies form the cornerstone of successful biofuel operations. While chapter 6 covers performance monitoring in depth, this section focuses on aligning preventive maintenance with overall risk reduction.

## Key strategies include:

- Scheduling maintenance based on manufacturer and operational history

- Using failure mode and effect analysis (FMEA) to identify vulnerable points

- Establishing maintenance KPIs (e.g., mean time between failures)

- Creating SOPs for high-risk equipment and ensuring parts availability

- Training and skill development for maintenance personnel

- Regular calibration and verification of monitoring systems

- Maintenance cost tracking and optimization

Baseline maintenance schedules, periodic inspections and planned redundancy are essential tools for managing production disruption risks even in the absence of real-time diagnostics.

Predictive maintenance systems analyze patterns in performance data to spot potential issues early, enabling maintenance teams to address problems before they lead to significant disruptions. For example, changes in equipment vibration patterns or thermal signatures can serve as indicators of developing mechanical issues that require attention.

Throughout this chapter, we have delved into the intricate relationship between technical challenges, operational vulnerabilities and the strategies required to effectively address them.

The case example and frameworks presented demonstrate that successful risk mitigation in biofuel operations is built on three fundamental pillars: robust technical assessment systems, comprehensive quality control protocols and adaptable emergency response procedures. When these elements are properly integrated, they establish a resilient operational framework capable of tackling current challenges and emerging risks in the rapidly evolving biofuel sector.

Implementing multi-layered risk management approaches and seamlessly integrating advanced feedstock analysis with real-time monitoring systems has consistently proven particularly effective in mitigating operational risks.

**Multi-Layered Risk Management Framework**

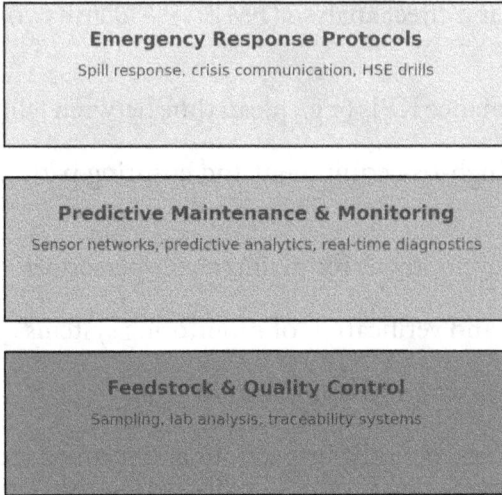

**Emergency Response Protocols**
Spill response, crisis communication, HSE drills

**Predictive Maintenance & Monitoring**
Sensor networks, predictive analytics, real-time diagnostics

**Feedstock & Quality Control**
Sampling, lab analysis, traceability systems

*\* Visual representation of layered risk management practices in industrial biofuel operations*

Looking forward, advances in monitoring technologies, data analytics and process control systems will likely continue playing a pivotal role in shaping the evolution of risk management in biofuel operations.

As we transition to the next chapter on environmental impact analysis, we build upon the understanding that effective risk management serves as a crucial enabler for sustainable biofuel operations.

# Key Takeaways

- **Risk management in using biofuel must be multi-layered**, addressing technical, operational and environmental risks through integrated systems that span feedstock quality, process control and emergency response.

- **Real-time monitoring and predictive analytics**—especially for feedstock variability and equipment performance—enable early detection of risks, supporting proactive rather than reactive interventions.

- **Comprehensive quality control systems** are essential and must extend across the entire value chain, combining automated tools with human oversight to ensure consistent product standards.

- **Emergency response and contingency planning** require regularly updated protocols, multi-supplier strategies and simulation exercises to maintain operational continuity during disruptions.

- **Preventive maintenance frameworks** powered by advanced sensors and AI-driven diagnostics reduce downtime, optimize cost efficiency and extend equipment life, strengthening overall system resilience.

Chapter 8

# Environmental Impact Analysis

*Measuring and Optimizing Sustainability Outcomes*

To pursue sustainable biofuel solutions effectively, a rigorous environmental impact analysis is essential. This analysis must go beyond simple carbon accounting to consider a wide range of ecological factors. As organizations aim to enhance their sustainability efforts, the ability to calculate, analyze and enhance environmental performance has become a crucial element for the successful implementation of biofuels. As we further explore environmental performance metrics, the complexity of measuring and optimizing emission outcomes becomes more evident. The shift to biofuels introduces specific challenges in quantifying environmental benefits while maintaining operational efficiency. Achieving this delicate balance necessitates advanced monitoring systems and comprehensive analytical frameworks capable of capturing both direct and indirect environmental impacts throughout the entire value chain.

When assessing the environmental impact of transitioning heavy machinery fleets to biofuels, traditional assessment methods fall short in capturing the complete scope of environmental changes. A breakthrough occurred with the development of a comprehensive system that monitors emissions, soil quality, water usage and biodiversity indicators. By integrating real-time environmental monitoring with operational data, organizations can enhance biofuel efficiency, safeguard the environment and maintain productivity.

Adaptive management now incorporates real-time environmental monitoring to adjust operations, enhancing environmental performance and often yielding significant cost savings through optimized resource utilization. However, true environmental impact analysis goes beyond measurement; it requires dynamic improvement systems.

This chapter delves into the methodologies and frameworks essential for conducting thorough environmental impact analyses in biofuel operations. We will explore various approaches to lifecycle assessment, investigate key performance indicators for emission reduction measurement, and discuss strategies for implementing effective monitoring systems. Special focus will be placed on integrating advanced monitoring technologies and data analytics to optimize environmental performance while preserving operational efficiency.

The challenge of measuring and optimizing emission outcomes in biofuel operations represents a critical frontier in the industry's development, necessitating sophisticated technical solutions and a fundamental shift in our approach to environmental impact assessment. By meticulously examining best practices and emerging methodologies, we can formulate more effective strategies to ensure that biofuel adoption upholds the promise of environmental sustainability.

## Lifecycle Assessment Methodologies and Environmental Impact Metrics

Lifecycle Assessment (LCA) methodologies are the cornerstone for evaluating the environmental impact of biofuel systems. Moreover, they provide a comprehensive framework for measuring emission outcomes across the entire value chain. Additionally, these methodologies examine environmental impacts from feedstock cultivation through processing, distribution and final use. Consequently, organizations can make data-driven decisions about their biofuel strategies.

**A robust LCA framework typically encompasses several key environmental impact metrics:**

- Greenhouse Gas Emissions ($CO_2$, $CH_4$, $N_2O$)

- Water Consumption and Quality Impact

- Land Use Change Effects

- Soil Quality and Biodiversity Impact

- Energy Return on Investment (EROI)

- Waste Generation and Management

Implementing standardized Life Cycle Assessment (LCA) methodologies requires careful consideration of system boundaries and allocation methods, particularly when evaluating the environmental effects of biofuels from farm waste. It's important to separate the impact of food crops from the effects of the biofuel crops to make informed decisions. The allocation decision can significantly influence the final assessment results and subsequent decision-making processes.

Environmental impact metrics must be both comprehensive and practical to enable organizations to track progress while maintaining operational efficiency. Key performance indicators (KPIs) should align with global sustainability frameworks while remaining relevant to specific operational contexts, often including direct measurements like carbon intensity per unit of fuel produced and indirect indicators such as changes in local biodiversity or soil organic carbon content.

Advanced monitoring systems are crucial for collecting and analyzing environmental impact data effectively. Modern LCA platforms integrate real-time monitoring capabilities with sophisticated analytics tools, enabling organizations to track environmental performance continuously rather than relying on periodic assessments. This dynamic approach to environmental monitoring allows for rapid identification of optimization opportunities and immediate response to potential environmental concerns.

The selection of appropriate environmental impact metrics should reflect regulatory requirements and organizational emission reduction goals. While carbon emissions often dominate discussions of environmental impact, a holistic assessment must consider the full spectrum of ecological effects, including impacts on local ecosystems, water resources and air quality, as well as broader implications for global environmental systems.

One particularly practical approach involves developing a hierarchical framework of environmental impact metrics, where primary indicators track critical environmental outcomes and secondary metrics provide deeper insights into specific aspects of environmental performance. This structured approach ensures comprehensive coverage while maintaining a focus on the most significant environmental impacts.

Continuous improvement in environmental performance requires regular review and refinement of assessment methodologies. With new science and better monitoring, organizations need to update how they assess environmental impact to ensure that emission assessments remain relevant and effective over time.

Integrating life cycle assessment methodologies with operational decision-making processes represents a critical success factor in optimizing environmental performance. Furthermore, by establishing clear links between environmental metrics and operational parameters, organizations can develop more effective strategies for reducing environmental impact while maintaining operational efficiency. As a result, this integrated approach enables more informed decision-making about technology selection, process optimization and resource allocation.

## Sustainability Optimization Strategies and Performance Benchmarking

Effective sustainability requires a systematic approach that balances both environmental and operational needs. Consequently, organizations should set clear targets and consistently strive for improvement. Moreover, this section explores key approaches to optimizing emission reduction outcomes and establishing effective performance benchmarking systems.

**Key optimization strategies for enhancing sustainability performance include:**

- Process integration and heat recovery systems

- Feedstock diversification and quality management

- Waste minimization and circular economy approaches

- Energy efficiency optimization

- Water conservation and treatment systems

To ensure the successful implementation of these strategies, organizations must be supported by robust performance benchmarking frameworks. These frameworks enable organizations to measure progress and identify areas for improvement effectively. Effective benchmarking, in turn, requires establishing baseline performance metrics, setting realistic improvement targets and implementing systems for tracking and reporting progress against these targets.

A structured approach to emission reduction optimization is essential. It begins with a comprehensive baseline assessment, followed by target setting and implementation planning. To demonstrate early success, organizations should identify quick wins while also developing longer-term strategies for more complex optimization initiatives. This balanced approach helps maintain momentum while building support for more ambitious environmental goals.

Performance benchmarking plays a crucial role in driving continuous improvement in environmental outcomes. Organizations should establish internal and external benchmarking processes to compare performance against industry standards and best practices. This comparative analysis helps identify improvement opportunities and validates the effectiveness of optimization strategies.

**Key considerations for developing effective benchmarking systems include:**

- Selection of appropriate peer groups for comparison

- Development of standardized measurement protocols

- Implementation of data collection and validation processes

- Regular review and updating of benchmarking criteria

By integrating advanced monitoring and analytics capabilities, sustainability optimization efforts can be significantly enhanced. Real-time performance monitoring allows for the rapid identification of deviations from target performance levels and enables prompt corrective action. These systems should be designed to provide actionable insights that drive continuous improvement in environmental outcomes.

Successful optimization strategies often involve a combination of technical solutions and operational improvements. For instance, implementing advanced process control systems can optimize energy consumption and reduce emissions, while improved maintenance practices can extend equipment life and reduce waste generation. The key is to develop integrated approaches that simultaneously address multiple sustainability objectives.

Performance benchmarking should go beyond simple metric comparison to include an analysis of best practices and success factors. Organizations should establish mechanisms for sharing knowledge and lessons learned across different operational units and with industry peers where appropriate. This collaborative approach can accelerate the adoption of successful optimization strategies and drive industry-wide improvements in emission reduction performance.

To maintain progress toward environmental goals, optimizing strategies must be regularly reviewed and refined. Organizations should establish formal review processes to evaluate the effectiveness of implemented measures and identify opportunities for further improvement. This method ensures that emission reduction optimization stays current by constantly adapting to new situations.

It is essential that sustainability strategies align with an organization's objectives and resources. Various factors such as financial considerations, operational requirements and regulatory compliance obligations influence the selection and implementation of optimization measures. Successful strategies prioritize core environmental objectives by considering these factors.

## Environmental Monitoring Systems and Continuous Improvement Frameworks

Environmental monitoring systems serve as the foundation for effective sustainability management in biofuel operations. They not only provide real-time insights and data-driven decision support for continuous improvement initiatives, but also integrate multiple sensor technologies, data analytics platforms and automated reporting tools to offer a comprehensive view of environmental performance across the entire operation.

The implementation of effective environmental monitoring systems requires careful consideration of **several key components:**

- Continuous Emissions Monitoring Systems (CEMS)

- Water quality monitoring networks

- Soil health assessment tools

- Biodiversity impact tracking systems

- Resource consumption meters

- Waste management analytics

Modern environmental monitoring platforms harness cutting-edge technologies like IoT sensors, satellite imaging and artificial intelligence to offer unparalleled insight into environmental impacts. By detecting and addressing potential issues proactively, organizations can prevent significant problems and pinpoint areas for optimization and efficiency enhancement.

The integration of environmental monitoring data with operational control systems fosters powerful synergies for environmental optimization, establishing clear links between operational parameters and environmental outcomes. This enables organizations to devise more effective strategies for reducing environmental impact while upholding operational efficiency, allowing for real-time adjustments to process parameters based on environmental performance indicators.

Continuous improvement frameworks are instrumental in translating monitoring data into actionable insights and sustainable performance gains. Typically, following a structured approach, these frameworks facilitate the implementation of strategies for sustainable performance enhancement.

- Regular performance review cycles

- Root cause analysis of environmental incidents

- Systematic implementation of corrective actions

- Validation of improvement measures

- Documentation of lessons learned

The success of continuous improvement initiatives heavily relies on establishing clear accountability and responsibility for environmental performance throughout the organization. To achieve this, key strategies include developing appropriate training programs, establishing clear communication channels and implementing effective feedback mechanisms to identify and address improvement opportunities promptly.

Advanced analytics capabilities play a crucial role in enhancing the effectiveness of environmental monitoring systems. They enable predictive maintenance, early warning detection and optimization modeling. Machine learning algorithms further contribute by analyzing complex patterns in environmental data to identify potential issues before they manifest as problems, enabling proactive environmental management rather than reactive responses.

It is essential to consider regulatory compliance and stakeholder expectations when developing environmental monitoring. While compliance requirements often set minimum monitoring standards, leading organizations typically go beyond these to implement more extensive monitoring capabilities that support their environmental objectives and maintain their social license to operate.

By integrating environmental monitoring systems with broader sustainability management platforms, organizations can develop more holistic approaches to environmental performance optimization. This integrated approach supports better decision-making by providing comprehensive visibility of the environmental implications of operational choices and investment decisions.

Regularly reviewing and improving monitoring systems and frameworks helps organizations continuously improve their environmental performance. To keep monitoring and improvement processes effective, organizations need to adapt to new technologies and environmental understanding. Continuous improvement ensures that environmental systems remain relevant and effective for sustainability.

Setting clear goals is crucial for improving environmental performance. These metrics should align with organizational environmental goals while providing meaningful insights into operational performance. Regularly reviewing goals helps keep improvement projects on track and ensures that targets are challenging but attainable. As we conclude this examination of environmental impact analysis in biofuel operations, several critical insights emerge that will shape the future of sustainability measurement and optimization. This chapter provides organizations with a strong foundation for maximizing the environmental benefits of biofuels while maintaining operational excellence.

The transition from traditional environmental assessment methods to dynamic, real-time monitoring systems represents a significant evolution in how we approach sustainability optimization. These advanced systems, integrating IoT sensors, artificial intelligence and sophisticated analytics platforms, enable organizations to move beyond simple compliance monitoring to proactive environmental management. The mining case example presented earlier demonstrates how this evolution can drive both environmental improvements and operational efficiencies, creating a powerful synergy between environmental goals and business performance.

In the context of biofuel operations, the importance of lifecycle assessment methodologies cannot be overstated. By examining environmental impacts across the entire value chain, from feedstock production to end-use applications, organizations can develop more effective strategies for reducing their ecological footprint. Implementing a tiered system of environmental impact measurements helps manage complexity and prioritize key emission reduction goals.

Moreover, this chapter has highlighted the essential role of continuous improvement frameworks in driving sustained environmental performance gains. By combining environmental monitoring and control systems, organizations can make real-time improvements and quicker responses, enabling them to adapt rapidly to changing conditions while maintaining progress toward their emission reduction objectives.

Looking ahead, the continued evolution of monitoring technologies and analytical capabilities will further enhance our ability to measure and optimize environmental outcomes. Organizations embracing these advances while maintaining a balanced focus on both environmental and operational performance will be best positioned to succeed in the rapidly evolving biofuel landscape. The key lies in developing flexible, adaptive systems that can grow alongside technological capabilities and stakeholder expectations.

Ultimately, the success of environmental impact analysis in biofuel operations depends on more than just sophisticated monitoring systems and measurement frameworks; it requires a fundamental commitment to continuous improvement and a willingness to adapt strategies based on emerging data and insights. By maintaining this focus on dynamic optimization while leveraging advanced monitoring capabilities, organizations can maximize the environmental benefits of their biofuel initiatives while building more resilient and sustainable operations for the future.

Technological advancements and better environmental understanding will continue to shape the principles and methods in this chapter, highlighting how thorough environmental reviews are crucial in sustainable biofuel production and emphasizing that success in sustainable energy requires adaptable organizations utilizing these methods.

## Key Takeaways

- **Life Cycle Assessment (LCA)** is essential for evaluating the full environmental impact of biofuels, covering emissions, land use, water, biodiversity and energy return across the entire value chain—from feedstock to fuel use.

- **Real-time environmental monitoring systems** (using IoT, AI and advanced sensors) enable organizations to shift from static compliance models to dynamic, proactive environmental management.

- **Tiered environmental metrics**—primary indicators for core impacts and secondary ones for granular detail—help organizations prioritize sustainability goals without overwhelming complexity.

- **Continuous improvement frameworks** (e.g., regular performance reviews, root cause analysis, corrective actions) ensure environmental gains are maintained and refined over time.

- **Sustainability optimization** succeeds when technical strategies (e.g., process integration, feedstock diversification) are combined with strong benchmarking, workforce training and stakeholder-aligned performance goals.

# Chapter 9

# FUTURE TECHNOLOGIES

*Emerging Solutions, AI and Innovation Pathways*

Innovation is rapidly redefining the biofuel landscape. This chapter explores the most transformative frontiers—from AI-driven process optimization to engineered microbes that produce carbon-negative fuels.

The horizon of biofuel innovation stretches far beyond current technologies, potentially transformative solutions that could revolutionize how we produce and utilize renewable energy. New technologies like AI and biotechnology are creating exciting possibilities for sustainable fuel production, while powerful computers and advances in biology are opening up new opportunities in biofuel production.

Advanced algorithms now enable real-time optimization of complex biological processes, and synthetic biology techniques offer new pathways for designing more efficient fuel-producing organisms. This technological renaissance is particularly evident in the field of artificial intelligence applications for biofuel production.

**Case Study**
*The integration of AI-driven optimization systems initially met resistance from veteran operators skeptical of automated control methods. However, implementing machine learning algorithms capable of analyzing thousands of process variables in real time showed capabilities far beyond traditional manual oversight. The AI system identified subtle patterns in feedstock variations and processing conditions that had previously gone undetected, automatically adjusting parameters to optimize yield while reducing energy consumption. This success story highlights how emerging technologies can bridge the gap between current limitations and future possibilities in biofuel production.*

The landscape of biofuel innovation extends far beyond artificial intelligence, with biotechnology breakthroughs enabling the development of engineered microorganisms capable of more efficient fuel production. New techniques in chemistry and biology are revolutionizing fuel production, occurring as part of an interconnected web of technological advancement.

In this chapter, we will examine real-world examples where emerging tech has not only improved efficiency but also overcome long-standing barriers like feedstock variability and yield limits. This isn't just about promising breakthroughs—it's about integrating disruptive tools into real operations.

# Advanced Biotechnology and Synthetic Biology in Biofuel Production

The intersection of biotechnology and synthetic biology represents a promising frontier in biofuel production advancement, revolutionizing our approach to creating sustainable fuels. These cutting-edge fields offer unprecedented opportunities to optimize production processes and develop novel fuel molecules. Scientists have genetically engineered microbes to produce fuel more efficiently and use less energy.

### Benefits

- Conversion of diverse feedstocks (e.g., lignocellulosic waste, CO2)

- Higher product yields with fewer byproducts

- On-site customization of metabolic pathways for fuel targets

The ability to design and modify biological systems at the genetic level has opened new possibilities for creating optimized production strains that can operate under various conditions while maintaining consistent output.

Metabolic engineering has emerged as a powerful tool in the biofuel sector, allowing for the creation of synthetic pathways that can produce advanced biofuels with superior properties. By manipulating cellular metabolism, scientists have engineered microorganisms that can produce fuel molecules with higher energy density and better combustion characteristics than traditional biofuels. This method not only improves fuel quality but also enhances production efficiency, resulting in lower costs.

Integrating advanced biotechnology into biofuel production systems has led to significant breakthroughs in process optimization. For instance, the development of consolidated bio-processing (CBP) organisms has simplified the production process by combining multiple steps into a single operation. These engineered microorganisms can perform both biomass breakdown and fuel production, reducing operational complexity and production costs while increasing overall efficiency.

### Case Example
*LanzaTech, in collaboration with Danone, has developed a method to produce monoethylene glycol (MEG), a key component of PET, directly from captured carbon emissions. This process utilizes a proprietary engineered bacterium to ferment emissions from steel mills or gasified waste biomass into MEG, eliminating the need for an ethanol intermediate and simplifying the supply chain. The laboratory-scale proof of concept has been validated by external laboratories. This breakthrough has the potential to reduce the environmental impact of PET production across various sectors, including packaging and textiles. The consortium plans to scale up the technology with Danone's support.*

### Case Example
*Anellotech's Bio-TCat™ technology, developed in partnership with IFP Energies Nouvelles (IFPEN) and Axens, is now ready for commercialization. This process converts woody biomass into bio-based aromatics like p-xylene, toluene and benzene. Suntory utilized Bio-TCat™ p-xylene to produce 100% bio-based PET resin, successfully creating prototype beverage bottles. The achievement follows over 5,000 hours of pilot plant operation using pine wood feedstock. This milestone demonstrates the technology's potential to produce cost-competitive renewable chemicals, supporting the development of sustainable products.*

Synthetic biology approaches enable the development of novel biosensors and regulatory circuits that can monitor and control production processes in real time, detecting metabolic bottlenecks, adjusting production parameters and maintaining optimal conditions for fuel synthesis. This leads to more stable and efficient production processes. Incorporating these sophisticated control mechanisms at the cellular level represents a significant advance over traditional process control methods.

However, acknowledging the challenges of implementing these advanced biological systems in industrial settings is essential.

Combining synthetic biology with AI and advanced analytics will accelerate innovation, as advanced tools help researchers understand and improve complex biological systems using machine learning. Integrating computational and biological approaches creates new opportunities for developing more sustainable and efficient biofuel production systems.

## Challenges:

- Regulatory hurdles for engineered organisms

- Containment and contamination risks

- Strain stability over time

- Scale-up from lab to bioreactor

### Case Example
*LanzaTech engineered a microbe to convert industrial CO2 emissions directly into ethanol. The process has now been commercialized at sites in China and Belgium, reducing fossil carbon input and showcasing industrial-scale synthetic biology.*

The economic implications of these technological advances are significant. Specifically, as production processes become more efficient and scalable, biofuels' cost-competitiveness continues to improve. Moreover, this progress is particularly important for industries seeking to decarbonize their operations while maintaining operational efficiency. Consequently, the ability to produce advanced biofuels through optimized biological processes offers a promising pathway toward sustainable energy solutions that can compete effectively with traditional fossil fuels.

## Artificial Intelligence and Machine Learning Applications in Process Optimization

AI and machine learning technologies have revolutionized biofuel production, enhancing its efficiency. These advancements have also led to improvements in monitoring and controlling biofuel production processes. By utilizing advanced algorithms and data analytics, facilities can now achieve unparalleled precision in process control while reducing operational costs and environmental impact.

**Benefits**

- Real-time process optimization of chemical composition measurements

- Predictive maintenance

- Dynamic feedstock blending

- Energy efficiency through machine learning insights

Implementing AI-driven control systems involves careful consideration of existing operational frameworks and operator expertise. Success in this endeavor hinges on creating hybrid systems that blend AI's pattern recognition capabilities with human experience and judgment. Rather than replacing human decision-making, these systems should augment it, providing operators with enhanced insights and recommendations while still allowing them to maintain ultimate control over critical processes.

> "We used to build reactors first and hope biology would keep up. Now, with AI-powered metabolic design, we're flipping that model—designing microbial systems first and engineering environments around them. It's a paradigm shift." - Dr. Sanae Kobayashi, Bio-design specialist, Institute for Renewable Chemistry

Deep learning algorithms have demonstrated effectiveness in optimizing feedstock selection and preprocessing. By analyzing the chemical composition of different feedstock sources and predicting their potential yield and processing requirements, these systems enable facilities to fine-tune their input mixture for maximum efficiency. This capability becomes increasingly valuable as facilities handle diverse and variable feedstock sources.

AI and IoT sensors work in tandem to create robust monitoring systems for production, enabling the collection and analysis of data from every stage, from feedstock handling to final product storage. The insights gathered drive continuous improvement initiatives and support data-driven decision-making at all operational levels.

However, implementing AI systems in biofuel production facilities poses unique challenges that need careful consideration.

## Challenges

- High upfront cost for systems integration

- Data quality and legacy infrastructure limitations

- Sensor reliability

- Workforce reskilling needs

- Cybersecurity risks

The future of AI in biofuel production holds great promise, with new algorithms and computing capabilities constantly evolving. Quantum computing applications may soon enable even more sophisticated optimization models, while advances in edge computing could enhance real-time processing capabilities. These developments indicate that the potential for AI-driven optimization in biofuel production is far from fully realized.

**Case Example**
*A northern European bio-refinery deployed AI to manage heat exchangers and fermentation tanks. By training the model on three years of sensor data, the AI system was able to reduce energy consumption and increase ethanol yield within four months. Operator training and staged integration were crucial for success.*

Implementing AI systems in biofuel production can yield substantial economic benefits. While the initial investment costs may be significant, the long-term advantages, such as improved efficiency, reduced waste and enhanced product quality, typically justify the expenditure. As these technologies become more widespread, implementation costs decrease while capabilities expand.

Successful AI implementation relies on comprehensive training and workforce development. Operators and technical staff need proper training to effectively utilize these new tools and interpret their outputs. This investment in human capital ensures that facilities can fully leverage AI systems' capabilities while maintaining safe and reliable operations.

By combining AI with biotechnology and process improvements, new opportunities for biofuel production can be created. This integration could significantly enhance the efficiency and sustainability of renewable fuel production.

**Implementation Caveat**: AI systems require consistent, high-quality data and specialized training—without these, predictive models often underperform in industrial biofuel settings.

# Next-Generation Catalysis and Processing Technologies

Advancements in catalysis and processing play a crucial role in driving down the cost and enhancing the efficiency of biofuels. These breakthroughs are revolutionizing the way we convert biomass into sustainable fuels, providing unparalleled opportunities for streamlining processes and maximizing output. By creating innovative catalytic systems and employing cutting-edge processing techniques, we can establish more streamlined conversion pathways that not only decrease energy consumption but also lessen the environmental impact.

> "One of the greatest misconceptions is that technological performance alone determines success. In my experience, process efficiency only matters if you can replicate it under variable real-world conditions — across feedstock types, humidity levels, and reactor loads." - Dr. Naveen Rao, Process Engineer

**Benefits**

- More efficient biomass conversion pathways

- Lower reaction temperatures and pressures

- Fewer downstream purification steps

One groundbreaking technology in biofuel production is selective catalysis, which grants greater precision in managing reaction paths and product distributions. Particularly noteworthy are the advanced heterogeneous catalysts, especially those utilizing nanoscale materials and engineered surface properties, which exhibit exceptional efficiency and selectivity enhancements. These catalysts are capable of operating at reduced temperatures and pressures compared to traditional methods, resulting in significant energy savings without compromising yield rates.

A promising trend in this domain is the integration of multifunctional catalysts that can execute multiple conversion steps concurrently, simplifying and cost-reducing production processes. By combining reactions, efficiency is boosted, unwanted byproducts are minimized and product quality is improved.

### Case Example

*Clariant has commenced production of cellulosic ethanol at its sunliquid® plant in Podari, Romania. The facility processes approximately 250,000 tons of locally sourced agricultural residues to produce around 50,000 tons of ethanol annually. The entire output is contracted to Shell under a multi-year agreement. This advanced biofuel can be used for fuel blending and has applications in sustainable aviation fuel and bio-based chemicals. The project supports Clariant's strategy to commercialize its sunliquid® technology and contributes to reducing greenhouse gas emissions in the transport sector.*

Advancements in biofuel production technology include process intensification through innovative reactor designs, which offer enhanced control over reaction conditions and improved heat and mass transfer efficiency. Micro-reactor systems and continuous flow processing methods enable more precise temperature control and better mixing characteristics, leading to more consistent product quality and reduced energy consumption. Additionally, the development of membrane-based separation technologies has revolutionized downstream processing in biofuel production. These advanced membrane materials and configurations facilitate more efficient separation of products from reaction mixtures, reducing energy requirements for purification and improving product recovery rates, especially in water removal and product concentration steps that address key challenges in biofuel processing efficiency.

However, the implementation of these advanced technologies requires careful consideration of scalability and economic factors. While laboratory demonstrations often show promising results, scaling these processes to commercial production volumes presents significant engineering challenges that require both technical skill and practical application to succeed.

## Challenges

- Catalyst deactivation and regeneration

- Scale-up from pilot to full-scale plant

- Integration with existing heat and flow systems

By combining better controls and catalysts, real-time reaction optimization can be achieved. Smart catalyst systems, which are equipped with embedded sensors, can provide continuous feedback on reaction progress and catalyst performance. This enables dynamic adjustments to maintain optimal operating conditions. This level of control not only helps maximize catalyst lifetime but also ensures consistent product quality.

In the future, the convergence of advanced catalysis with other emerging technologies holds promise for further improvements in biofuel production efficiency. For instance, combining engineered catalysts with synthetic biology approaches opens up new possibilities for directly converting biomass into advanced biofuels. These hybrid approaches have the potential to overcome current limitations in conversion efficiency while reducing processing complexity.

The economic implications of these technological advances are significant for the biofuel industry, as advances in processing and catalysts are making biofuels more competitive. This progress is crucial for achieving broader market adoption and supporting the transition to sustainable energy sources across various industrial sectors.

Developing robust catalyst regeneration and recycling protocols has become increasingly important as the industry moves toward more sustainable practices. Advanced regeneration techniques help extend catalyst lifetime while reducing waste and replacement costs, aligning with circular economy principles and improving the overall economics of biofuel production.

## Strategic Integration Roadmap

| Phase | Focus Area | Key Action Items |
|---|---|---|
| Pilot Integration | Single-unit optimization | Sensor installation, operator co-training |
| Modular Scaling | Line-wide deployment | Digital twins, strain biocontainment protocols |
| Systemic Upgrade | Cross-functional integration | Catalyst retrofits, AI control rooms, ESG metrics |

Cost Estimate (Range)

- AI Process Suite: $500K–$2M

- Synthetic Biology Scale-up: $2M–$10M per strain

- Catalysis Redesign and Retrofitting: $1M–$5M per system

## Technology Integration Challenges and Enablers

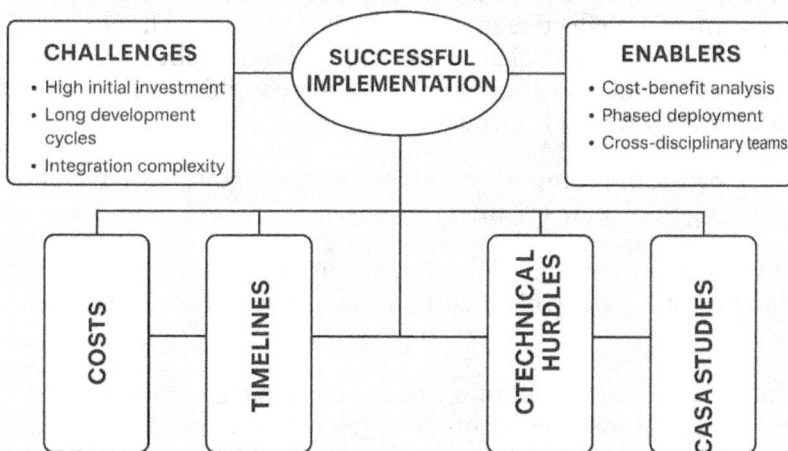

**CHALLENGES**
- High initial investment
- Long development cycles
- Integration complexity

**SUCCESSFUL IMPLEMENTATION**

**ENABLERS**
- Cost-benefit analysis
- Phased deployment
- Cross-disciplinary teams

COSTS

TIMELINES

CTECHNICAL HURDLES

CASA STUDIES

As we conclude our exploration of emerging technologies and innovation pathways in the biofuel sector, it becomes clear that we stand at the threshold of a potentially transformative era in sustainable energy production, where combining AI, advanced biotech and better catalysts is opening up new possibilities for biofuels. These technological advances, while promising, require careful consideration of practical implementation challenges and economic viability.

Biofuel production efficiency is significantly improved by AI optimization systems, with success in fully implementing machine learning algorithms capable of analyzing thousands of process variables in real time, representing a significant step forward in achieving environmental sustainability and economic viability. However, technology alone is not enough; successful adoption needs careful integration with people and existing processes.

Advanced biotechnology and synthetic biology have opened new frontiers in biofuel production, enabling the development of enhanced microorganisms and more efficient conversion processes. Genetically engineering biological systems allows for optimized fuel production with minimal resource use, but successfully using these technologies commercially requires careful consideration of safety and practical issues.

Advancements in catalysts and processes are resulting in improved conversion rates and product quality, facilitated by enhanced catalysts and reactor designs that enable more cost-effective and intensified processes.

Nevertheless, it is essential to carefully consider the trade-offs between the advantages of these innovations and the actual costs and operational requirements. The effective implementation of new technologies hinges on addressing several key challenges. Establishing a robust infrastructure and support systems capable of managing these sophisticated technologies on a commercial scale is imperative.

Secondly, providing comprehensive training programs to prepare workers for operating these new systems is crucial.

Thirdly, the adoption of flexible regulations is necessary to accommodate emerging technologies while maintaining safety and environmental standards.

The future of biofuel technology lies in the strategic integration of various innovative approaches, rather than relying on singular breakthroughs. Success in this field is contingent upon a pragmatic evaluation of new technologies, balancing their potential with the practical needs of the industrial sector. Moving forward, the emphasis should be on developing solutions that are advanced, economically viable and environmentally sustainable.

---

🏴 *Reality Check: Emerging Tech, Real-World Hurdles*

*Despite their promise, advanced technologies like synthetic biology and AI-integrated control systems face non-trivial implementation barriers. Industrial-scale bioreactors must address contamination control, regulatory biosafety standards and strain stability across variable conditions. Similarly, AI-based optimization tools are only as good as the data quality; many facilities lack the instrumentation to generate clean, consistent input data streams.*

*Key Watchpoints:*
*\* Strain reproducibility and containment safeguards in bio-engineered microbes*
*\* Cost and training required for AI/ML integration*
*\* High upfront investment vs. long-term efficiency gains*

---

# Key Takeaways

- **AI and machine learning** are revolutionizing biofuel production by enabling real-time process optimization, predictive maintenance and intelligent feedstock management—significantly reducing waste and energy consumption.

- **Synthetic biology and metabolic engineering** have produced genetically modified microorganisms that can process multiple feedstocks and generate high-quality fuels with improved efficiency and resilience.

- **Next-generation catalysts** and advanced reactor designs (e.g., micro-reactors, membrane separation) are streamlining biomass conversion, lowering operational costs and improving yield and product quality.

- **Integration is key**: The convergence of AI, biotechnology and catalysis is more powerful than any single breakthrough—hybrid systems are driving potentially transformative improvements in sustainability and scalability.

- **Practical implementation matters**: Technological success depends on human training, infrastructure readiness, regulatory flexibility and industrial scalability—innovation must align with real-world constraints.

Chapter 10

# STRATEGIC PLANNING
*Building Resilient Biofuel Business Models*

Developing resilient biofuel business models requires a strategic approach that transcends traditional energy sector paradigms. This involves combining innovative operational frameworks with robust risk management systems. In an industry characterized by rapid technological advancement and evolving market dynamics, success depends on building adaptable business structures that can navigate both current challenges and future opportunities. At the heart of this resilience lies the ability to anticipate and adapt to market shifts while maintaining operational stability. Key factors such as new technology, sustainability and smart risk management play crucial roles in ensuring long-term biofuel success.

> **Case Example**
> *A mining operation faced significant volatility in its conventional fuel supply chain. The existing linear business model, which prioritized production efficiency above all else, proved inadequate in managing supply chain disruptions and adapting to shifting regulatory requirements. Through strategic analysis and planning, a circular business model was developed. This model integrated feedstock recycling, waste minimization and flexible production capabilities to mitigate risks. Despite initial stakeholder resistance and technical hurdles in process adaptation, this adaptive model demonstrated that profitability could be sustained even during market downturns while significantly reducing environmental impact.*

This transformation underscores a fundamental truth in the biofuel sector: building resilient business models extends beyond financial planning. It requires creating dynamic systems that can thrive in an increasingly complex and unpredictable market environment. Succeeding in this endeavor necessitates considering the environmental, social and economic impacts as integral components of the business strategy.

This chapter explores the essential components of building resilient biofuel business models, examining how organizations can develop adaptive strategies that respond to market dynamics while maintaining operational excellence. It discusses innovation's impact on sustainable competitive advantages and risk management in a changing regulatory landscape, using practical examples and strategic frameworks to demonstrate how companies can plan for long-term growth within the biofuel sector.

Resilient biofuel businesses require balancing innovation and stability, along with efficient and adaptable risk management. As we navigate these complementary forces, we will discover how organizations can create sustainable value while contributing to the broader transition toward renewable energy sources.

## Business Model Innovation and Value Chain Optimization

Innovating the business model in the biofuel sector involves reimagining additional value chain structures to establish sustainable competitive advantages and enhance operational efficiency. To create resilient biofuel operations, organizations must navigate the challenges associated with transitioning from linear to circular value chains. This shift necessitates a thoughtful approach to various aspects, including feedstock sourcing, production processes, distribution networks and end-user applications.

An essential element of successful business model innovation is the identification and utilization of strategic partnerships throughout the value chain. These partnerships play a vital role in addressing common obstacles like feedstock supply volatility, infrastructure limitations and market access barriers.

For example, forming enduring agreements with agricultural producers can ensure a steady supply of feedstock while supporting rural economic growth. Similarly, collaborations with logistics providers and end-users can optimize distribution networks and maintain a consistent demand for biofuel products.

Optimizing biofuel production involves enhancing efficiency and sustainability through key strategies such as turning waste into value, using resources wisely and implementing closed-loop systems. Organizations need to carefully balance these factors while also maintaining operational flexibility to adjust to evolving market conditions.

The integration of digital technologies is essential in optimizing biofuel value chains, with advanced analytics, Internet of Things (IoT) sensors and blockchain solutions playing a crucial role. These technologies enable real-time monitoring and optimization of production processes, supply chain operations, and quality control measures, providing valuable insights and improved tracking throughout the entire process.

**Key elements of successful business model innovation include:**

- Strategic partnership development across the value chain

- Integration of circular economy principles

- Implementation of digital monitoring and optimization systems

- Development of flexible production capabilities

- Creation of sustainable competitive advantages

> *"Biofuels are no longer fringe bets—they're strategic hedges against regulatory uncertainty and energy volatility."*
>
> *— Anya Becker, ESG Investor & Former Energy Market Analyst*

Improving value chains involves systematically finding and fixing problems to create more value. Key actions include conducting regular assessments of operational processes, supplier relationships and customer needs to identify areas for improvement. Organizations must also consider the impact of regulatory changes, technological advancements and market trends on their value chain structure.

Risk management plays a vital role in business model innovation and value chain optimization, requiring organizations to develop robust frameworks for identifying, assessing and mitigating risks across their operations. These frameworks should address both operational risks, such as supply chain disruptions and quality control issues and strategic risks related to market dynamics and regulatory changes.

Innovation in biofuels depends on creating valuable solutions to common challenges, which may involve developing new revenue streams from waste products, implementing creative financing mechanisms, or establishing novel partnership structures that share risks and rewards among stakeholders.

**Critical success factors for value chain optimization include:**

- Robust quality control systems

- Efficient logistics and distribution networks

- Strong supplier relationships

- Effective waste management and resource recovery

- Continuous improvement processes

Organizations must also consider the role of stakeholder engagement in business model innovation. Key aspects include maintaining open communication channels with suppliers, customers, regulators and local communities to ensure alignment of interests and support for new initiatives. Successful implementation often requires building trust and demonstrating value creation for all stakeholders involved in the value chain.

Changes in technology, markets and regulations are reshaping the biofuel industry. Therefore, organizations that can effectively combine these elements while maintaining operational excellence and stakeholder support are best positioned for long-term success in the industry. It is crucial to note that success requires constant improvement and adapting to change.

# Strategic Risk Assessment and Adaptive Management Frameworks

Strategic risk assessment and adaptive management frameworks are crucial for successful biofuel operations in today's complex business environment. These frameworks need to be flexible and cover various risks, including operational, financial, regulatory and environmental ones. To effectively implement comprehensive risk assessment protocols, organizations should take a systematic approach that combines quantitative analysis with qualitative insights.

## Integrated Risk Management Zones

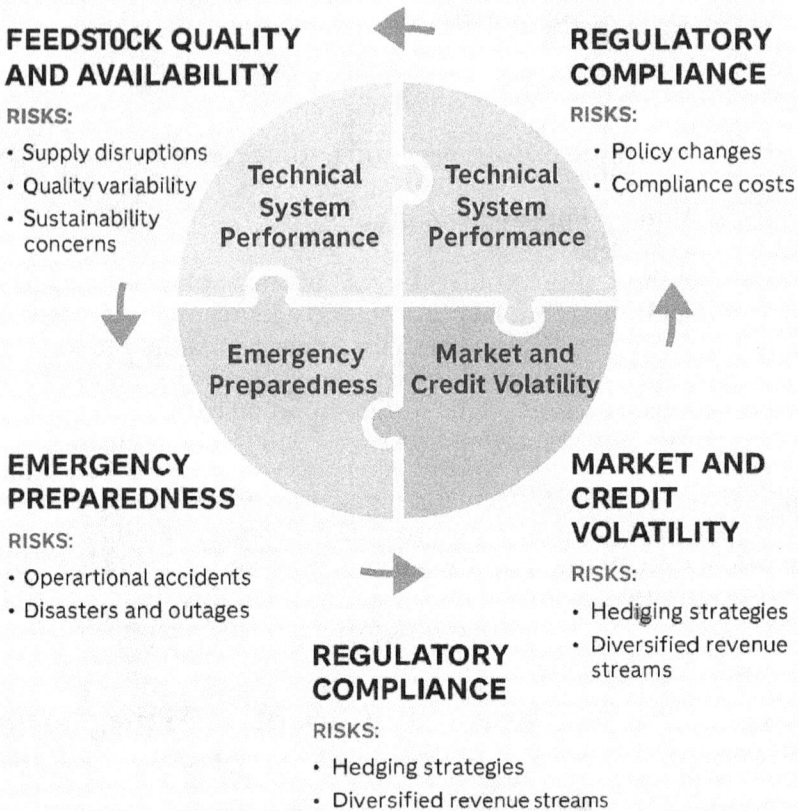

**FEEDSTOCK QUALITY AND AVAILABILITY**

RISKS:
- Supply disruptions
- Quality variability
- Sustainability concerns

Technical System Performance

**REGULATORY COMPLIANCE**

RISKS:
- Policy changes
- Compliance costs

Technical System Performance

Emergency Preparedness

Market and Credit Volatility

**EMERGENCY PREPAREDNESS**

RISKS:
- Operartional accidents
- Disasters and outages

**MARKET AND CREDIT VOLATILITY**

RISKS:
- Hedging strategies
- Diversified revenue streams

**REGULATORY COMPLIANCE**

RISKS:
- Hedging strategies
- Diversified revenue streams

When assessing risks, it is important for organizations to utilize clear metrics that take into account factors like supply chain reliability, production, market demand and regulatory compliance. This assessment process should be continuous and iterative, enabling organizations to refine their strategies based on operational experience and evolving market conditions.

### Key components of strategic risk assessment include:

- Systematic identification of potential risks and vulnerabilities

- Quantitative analysis of risk probability and impact

- Development of risk mitigation strategies

- Regular review and updating of risk assessments

- Integration with operational decision-making processes

Adaptive management frameworks build upon risk assessments by establishing flexible response mechanisms that can adjust to changing circumstances. These frameworks should incorporate clear decision-making protocols, communication channels and feedback loops to enable rapid response to both challenges and opportunities. The goal is to create resilient systems that maintain operational stability while adapting to evolving market conditions and regulatory requirements.

Consider the experience of a biofuel production facility that implemented a comprehensive risk assessment and adaptive management system. Initially focused primarily on operational risks, the facility expanded its framework to include broader strategic considerations such as market dynamics, regulatory changes and stakeholder expectations. This holistic approach enabled the facility to identify and address potential vulnerabilities before they materialized into significant problems, while also positioning it to capitalize on emerging opportunities.

Real-time monitoring of key indicators and risks is crucial for effective adaptation. Additionally, key components include implementing advanced analytics capabilities to process operational data, market intelligence and regulatory updates. Consequently, clear procedures are needed for handling and resolving issues that surpass risk limits.

**Essential elements of adaptive management frameworks include:**

- Real-time monitoring and reporting systems

- Clear decision-making hierarchies and protocols

- Flexible response mechanisms

- Regular testing and updating of contingency plans

- Integration with strategic planning processes

To ensure the success of these frameworks, a culture of risk awareness and adaptability is essential. This culture can be cultivated through regular training and communication to ensure that all stakeholders understand their roles in risk management and the importance of maintaining operational flexibility. Leadership plays a crucial role in demonstrating commitment to these principles by providing consistent support and allocating resources effectively.

Furthermore, organizations need to recognize the interconnected nature of risks within the biofuel sector. Changes in one area, such as feedstock availability, can have ripple effects across operations, market positioning and financial performance. Therefore, effective risk assessment frameworks should consider these relationships and incorporate scenario planning to evaluate potential outcomes under different conditions.

Developing adaptive management capabilities requires significant investment in both systems and people. Organizations must weigh the costs of implementing comprehensive risk management frameworks against the potential benefits of enhanced operational resilience and strategic flexibility. This balancing act involves investing in technology, training, and organizational development to support effective risk management, to maintain the effectiveness of risk assessment and adaptive management frameworks. It is essential for organizations to regularly review and update them. This can be achieved by establishing formal processes for evaluating system performance and incorporating lessons learned from operational experience. When significant events occur, conducting post-incident analyses can provide valuable insights that can be used to enhance future risk management capabilities.

By integrating strategic risk assessment and adaptive management frameworks with broader business planning processes, organizations can ensure that risk considerations influence strategic decision-making at all levels. This alignment helps organizations stay focused on long-term objectives while effectively managing short-term risks and operational challenges. In the biofuel sector, success increasingly hinges on developing these capabilities and maintaining the flexibility to adapt to changing market conditions while effectively managing risks.

## Market Positioning and Competitive Advantage Strategies

To achieve success in the biofuel industry, it is crucial for organizations to have a deep understanding of market trends, technology and needs. This understanding allows them to carefully assess their core competencies and market opportunities, enabling the development of sustainable competitive positions that generate lasting value for stakeholders and contribute to the broader energy transition.

A key aspect of successful market positioning involves identifying specific market segments that align with an organization's capabilities and customer requirements. This entails conducting detailed market analyses to gain insights into customer pain points, competitive dynamics and potential growth opportunities. For example, some organizations may choose to focus on serving heavy industry customers with specialized biofuel blends, while others may target transportation fleets looking to reduce their carbon footprint.

**Case Example**
*Consider the case of a biofuel producer that initially faced challenges in standing out in a competitive market. Through a thorough analysis of customer needs and operational capabilities, the producer discovered an opportunity to specialize in producing customized biofuel blends for industrial applications. By concentrating on a niche market segment and developing specialized processes, the producer gained a unique advantage over competitors.*

**Key elements of effective market positioning include:**

- Clear value proposition development

- Target market identification and segmentation

- Competitive analysis and differentiation strategy

- Customer relationship management

- Brand development and communication

Developing sustainable competitive advantages requires organizations to build and maintain capabilities that competitors cannot easily replicate. This success depends on leveraging new technology, establishing strong supply chains and offering innovative services. The key strategy is to focus on areas where the organization can create meaningful differentiation that customers value and are willing to pay for.

Furthermore, organizations should ensure that their plans align with industry trends and regulations. Key considerations include evaluating the impact of factors such as carbon pricing, renewable fuel standards and sustainability requirements on their competitive position in the long run.

Successful organizations often proactively position themselves ahead of regulatory changes, transforming compliance requirements into competitive advantages.

**Sources of competitive advantage in the biofuel sector may include:**

- Technological innovation and process efficiency

- Strategic partnerships and supply chain optimization

- Superior quality control and product consistency

- Regulatory compliance and certification

- Customer service and technical support capabilities

Developing competitive advantages often requires significant investment in tangible and intangible assets. Organizations must carefully evaluate the return on these investments while considering how they contribute to the long-term market position. This involves R&D, infrastructure improvements, workforce development and stakeholder engagement.

Maintaining competitive advantages in the dynamic biofuel market requires continuous innovation and adaptation. Companies should routinely evaluate their market position and competitive advantages to ensure long-term success, which includes monitoring changes in customer needs, competitive dynamics and technological capabilities that might affect their market position.

Effective communication of competitive advantages is crucial for market success. Companies must use distinct messaging that appeals to their target audience, with key strategies including articulating the unique value proposition of their products and services while building credibility through demonstrated performance and customer testimonials.

Strategic partnerships play a crucial role in creating competitive advantages for organizations. By forming alliances with complementary businesses, research institutions or industry groups, organizations can enhance their market position and gain access to new technologies, markets and capabilities that strengthen their competitive positions. These partnerships also allow for the sharing of risks and resources.

Long-term commitment and continuous improvement are essential for achieving market success, as companies must balance immediate goals with future investments to maintain competitiveness. It is important for companies to maintain flexibility to adapt to changing market conditions while staying true to their core strategic objectives.

As we wrap up our examination of strategic planning and resilient business models in the biofuel sector, several critical themes emerge that will shape the future of sustainable energy adoption. Success in the energy industry requires a smart approach to business models that goes beyond traditional methods, taking into account technology, markets and regulations. Throughout this chapter, we have explored how organizations can build adaptive frameworks that respond to market volatility while ensuring operational stability.

The mining case example presented demonstrates that profitable businesses can also be environmentally responsible. Achieving this transformation requires careful attention to supply chain optimization, stakeholder engagement and risk management—elements that form the foundation of resilient biofuel enterprises.

To effectively plan for biofuels, risk assessment and flexible management are necessary. The successful implementation of comprehensive monitoring systems and feedback loops enables organizations to identify and address potential vulnerabilities while capitalizing on emerging opportunities. Operational resilience to market and regulatory changes is achieved through flexible responses and effective risk management.

Biofuel competitiveness requires balancing innovation with stability. Organizations must carefully evaluate their core competencies and market opportunities while building sustainable value propositions that resonate with stakeholders. Success in this area demands continuous adaptation and refinement of business strategies supported by robust systems for monitoring performance and managing risk.

In order to grow sustainably, the biofuel sector must prioritize flexibility and build strong foundations. This includes integrating advanced technologies, circular economy principles, and stakeholder-centric approaches to develop business models that can thrive in a complex operating environment. Organizations that can effectively combine these elements while maintaining operational excellence will be best positioned to lead the transition toward sustainable energy solutions.

However, creating resilient biofuel businesses is an ongoing challenge that requires flexibility and responsiveness to market changes and stakeholder needs. Success in this endeavor demands a holistic approach that considers environmental impact, social responsibility and economic viability, as highlighted throughout this chapter. By embracing these principles and maintaining adaptability in the face of change, organizations can create lasting value while contributing to the broader energy transition.

The future of biofuel businesses hinges on technology, regulations and market demands. Organizations that can effectively navigate these forces while upholding operational excellence and stakeholder support will define the next chapter in sustainable energy development. Moving forward, the lessons and frameworks discussed in this chapter will serve as valuable guides for building resilient, sustainable businesses in the evolving biofuel landscape.

# Key Takeaways

- **Resilient biofuel business models prioritize adaptability** by integrating circular economy principles, flexible production capabilities, and real-time risk management frameworks to withstand market volatility and regulatory shifts.

- **Strategic innovation along the value chain**—including digital optimization tools, waste valorization and strong supply partnerships—enables competitive differentiation while driving operational efficiency and sustainability.

- **Risk assessment must be both strategic and adaptive**, incorporating real-time data, scenario planning and cross-functional communication to manage regulatory, supply chain, and technological uncertainties.

- **Market positioning relies on deep segmentation, targeted value propositions, and continuous differentiation** through customer-driven innovation, ESG performance and specialized product offerings.

- **Long-term competitiveness in biofuels is built through sustained innovation, stakeholder engagement, and strategic alliances**, which together create scalable, defensible advantages in a fast-evolving energy landscape.

# CONCLUSION

As we conclude our exploration of *The Biofuel Breakthrough* and its transformative potential for industrial nations, it becomes clear that the path toward sustainable energy is both challenging and promising. Throughout this book, we have examined how biofuels represent more than just an alternative energy source—they embody a fundamental shift in how we approach industrial energy consumption and environmental stewardship. The transition to sustainable, scalable biofuel solutions is no longer a theoretical aspiration—it is a critical necessity. But knowledge alone is not enough. Action must follow insight.

Throughout our journey across technological landscapes, market dynamics, policy mechanisms and implementation strategies, several critical insights have emerged that can propel industrial sectors toward a lower-carbon future. We have seen how AI, synthetic biology, supply chain digitization and new valuation models are reshaping the energy landscape. From pilot programs in Canadian rail yards to SAF deployments and bio-oil retrofits in the Nordic region, we've witnessed how success is possible when technology meets pragmatism, and bold thinking meets execution.

First, successful biofuel adoption requires a holistic approach that integrates technical expertise with strategic planning and stakeholder engagement. Case studies and examples presented throughout this book clearly demonstrate that even the most sophisticated technological solutions must be supported by robust implementation frameworks and careful consideration of local contexts.

Second, the role of sustainable and ethical supply chains has emerged as a crucial factor in the viability of biofuel initiatives. As we have seen, developing resilient procurement practices and implementing digital monitoring systems can significantly enhance operational efficiency while ensuring environmental compliance. For example, the blockchain-based tracking systems and local collection hubs discussed in earlier chapters illustrate how technology can bridge the gap between small-scale producers and industrial-scale operations.

Third, while financial considerations remain paramount, our analysis has shown that traditional valuation methods must evolve to capture the full spectrum of benefits offered by biofuel investments. In this regard, the hybrid valuation models presented in this book - incorporating both conventional metrics and environmental factors - provide a more comprehensive framework for assessing project viability in this emerging sector.

Moreover, keeping up with evolving regulations requires vigilance and flexibility. To illustrate, the integrated compliance matrices and multi-jurisdictional frameworks have explored practical tools for managing complex regulatory requirements while maintaining operational efficiency. Consequently, these approaches show that compliance can potentially be transformed from a burden into a strategic advantage.

Looking ahead, the future of biofuels appears increasingly bright, driven by technological innovation and growing environmental awareness. Specifically, new technologies like AI and advanced biotech may make biofuel production more efficient and sustainable. However, as we have emphasized throughout this book, successful implementation depends on both technological capability and the development of resilient business models and strategic planning frameworks.

Perhaps most importantly, switching to biofuels could revolutionize industrial energy, benefiting both businesses and the environment. In this context, the real-time monitoring systems, adaptive management frameworks and circular business models discussed in these pages illustrate how organizations can achieve operational excellence while contributing to global environmental goals.

At this critical juncture in the global energy transition, this book's insights and strategies provide a roadmap for organizations seeking to navigate the complexities of biofuel adoption. In summary, the success stories, lessons learned and practical frameworks provided here show that, while challenges exist, they can be overcome through careful planning, strategic implementation and an unwavering commitment to sustainability.

Ultimately, *The Biofuel Breakthrough* is not just about replacing one fuel source with another—it is about fundamentally transforming how we think about energy, emission reduction and industrial development. As we move forward, the principles and practices outlined in this book will continue to serve as guideposts for organizations and individuals committed to building a more sustainable future through the power of biofuels.

While the promise of biofuels is vast, it is equally important to acknowledge the roadblocks that remain. From fluctuating feedstock markets to infrastructure incompatibilities and fragmented global regulation, the path forward is far from frictionless. Many early pilots have stumbled due to underestimated costs, inconsistent quality control or insufficient training. However, these setbacks are not signs of failure—they are lessons that shape smarter strategies. By confronting these complexities head-on, the biofuel sector can evolve more resiliently and equitably.

At this pivotal moment in the global energy transition, **passive optimism must give way to informed action.** Whether you are a plant operator, a procurement manager, a policymaker or a citizen championing climate solutions, the path forward requires more than belief—it demands bold, pragmatic steps.

## How Can You Accelerate the Biofuel Transition?

**For Energy Professionals:**

- **Audit your current systems.** Use the diagnostic frameworks provided in chapter 6 to identify compatibility gaps and opportunities for early-stage adoption.

- **Champion pilot projects.** Start with limited-scope implementations. Use data to scale.

- **Build internal buy-in.** Train your teams, share KPIs and track outcomes.

- **Adopt adaptive planning.** Utilize digital twins, IoT sensors, and flexible contracts to future-proof your operations.

**For Policymakers:**

- **Create stable credit frameworks.** Support LCFS and RIN markets with multi-year policy certainty to spur private investment.

- **Fund infrastructure and R&D.** Support pre-commercial technologies that address feedstock logistics, distributed processing and SAF deployment.

- **Streamline permitting.** Implement clear, tech-neutral standards that accelerate project timelines.

- **Partner with industry.** Incentivize adoption through tax relief, green bonds and public-private demonstration pilots.

**For Sustainability Leaders and Investors:**

- **Push for hybrid valuation models** that include ESG and policy-adjusted metrics.

- **Demand transparency** in lifecycle emissions from suppliers and investment targets.

- **Back regional biofuel innovation hubs** and local supply chain integration projects.

**For Concerned Citizens and Sustainability Advocates:**

- **Support sustainable fuel policies.** Advocate for smarter, science-based renewable fuel standards in your region.

- **Ask companies how they fuel their operations.** Encourage biofuel usage in logistics, travel and transport sectors.

- **Educate your community.** Share what you've learned about feedstock sourcing, emissions reduction and the role of circular economies in decarbonization.

The biofuel revolution is not waiting for perfection—it is being built in real time by engineers, policymakers, financiers and citizens who refuse to stand still. Whether you're guiding fleets, shaping regulations or simply casting your vote, your role matters.

This is your moment to lead—not later, but now.

Let this book be your launchpad. Let your actions be the breakthrough.

# Leave A Review

*Thank You!*

If this book resonated with you or supported you in any way, you're warmly invited to leave a review on the platform where you purchased this book — reviews help other readers discover books like this.

If you'd like, you can also access the BIOFUEL READINESS PACK created to support and complement your reading experience.

*Scan Me: REVIEW and access BIOFUEL READINESS PACK*

# GLOSSARY

## A–C

**Advanced Biofuels**: Fuels derived from non-food biomass (e.g., agricultural waste, algae), typically offering greater environmental benefits than first-generation biofuels.

**Algae:** A diverse group of photosynthetic, aquatic organisms, ranging from microscopic single-celled organisms to large seaweeds.

**Anaerobic Digestion**: A biological process that breaks down organic material in the absence of oxygen, producing biogas and digestate.

**ASTM D6751**: Biodiesel standard in the U.S. for fuel quality

**B5:** 5% biodiesel, 95% petroleum diesel blend

**B20**: 20% biodiesel, 80% petroleum diesel blend

**B100:** 100% biodiesel

**Biodiesel (B100, B20)**: A renewable fuel made from vegetable oils or animal fats. B100 is 100% biodiesel; B20 is a 20% biodiesel blend with petroleum diesel.

**Bio-oil**: A crude-like fuel created via fast pyrolysis of biomass, typically requiring upgrading before use in engines.

**Biomass**: Organic material from plants or animals used as feedstock for biofuel production.

**Bioreactor Design:** Creating a controlled environment for biological processes like cell growth and product synthesis, ensuring optimal conditions for microorganisms or cells to perform their desired function.

**CapEx (Capital Expenditure)**: Upfront costs incurred when building or retrofitting facilities for biofuel processing or infrastructure.

**Carbon Intensity (CI)**: A measure of how much carbon dioxide is emitted per unit of energy produced, often expressed in grams $CO_2e$ per MJ.

**Catalyst Efficiency:** How effectively a catalyst transforms reactants into products in a chemical reaction.

**CBIO Decarbonization Credit:** An instrument adopted by RenovaBio Brazil as a tool to reach this target.

**Cellulosic Ethanol**: Ethanol produced from non-food plant materials like crop residues or switchgrass, requiring advanced pretreatment.

**Cloud Point**: The Temperature at which wax crystals begin to form in biodiesel

**CO2:** A clear gas composed of one atom of carbon (C) and two atoms of oxygen (O)

**Cyanobacteria:** A division of microorganisms that are related to the bacteria but are capable of photosynthesis

# D–L

**DPF (Diesel Particulate Filter): An** Emission control device that captures soot.

**Drop-in Fuel**: A biofuel that can directly replace petroleum fuels in existing engines and infrastructure with little or no modification.

**E10:** A blend containing 90% gasoline and 10% ethanol.

**E15:** A blend containing 15% ethanol and 85% gasoline.

**E85:** A blend containing 85% ethanol and 15% gasoline.

**Electrochemical Conversion:** The process of transforming electrical energy into chemical energy, or vice versa, through chemical reactions using electricity.

**EN 14214:** European standard for biodiesel fuel.

**Energy Density**: The amount of energy stored in a given system or fuel per unit volume or mass.

**Environmental NGO:** A non-governmental organization that focuses on environmentalism and works to address various environmental issues.

**Enzymatic Hydrolysis:** A process where enzymes break down large molecules, like proteins or carbohydrates, into smaller, simpler ones using water.

**EPA RFS Credits:** Specifically, Renewable Identification Numbers (RINs) are tradable credits generated by producers of qualifying renewable fuels.

**ESG (Environmental, Social, and Governance)**: Non-financial performance indicators are increasingly used in investment decision-making.

**Ethanol:** A clear, colorless liquid with a characteristic pleasant odor and burning taste.

**Feedstock**: Raw material used for biofuel production—e.g., corn, used cooking oil, wood chips.

**Flex-Fuel Vehicles (FFVs)**: Vehicles capable of running on gasoline or high-ethanol blends like E85.

**Fossil Fuel:** Naturally occurring, flammable materials formed from the remains of ancient organisms.

**GHGenius:** A spreadsheet-based Life Cycle Assessment (LCA) model primarily used to assess the energy and emissions associated with various transportation fuels, including conventional and alternative fuels.

**GHG Reduction Potential**: The extent to which a fuel or process lowers greenhouse gas emissions compared to fossil equivalents.

**Greenhouse Gas Emission (GHG)**: The release of gases into the atmosphere that trap heat, contributing to the greenhouse effect and global warming.

**(GREET) Greenhouse gases, Regulated Emissions, and Energy Use in Technologies**

**HEFA Diesel (Hydroprocessed Esters and Fatty Acids)**: A renewable diesel made by hydrotreating fats, oils and greases; chemically similar to petroleum diesel.

**Hydrolysis:** An important process in organic chemistry that involves the breakdown of molecules.

**Hydrotreated Vegetable Oil (HVO):** also known as renewable diesel or green diesel, is a biofuel produced from renewable resources like vegetable oil, animal fats or used cooking oil through a process called hydrotreating.

**Hygroscopic:** Tending to absorb moisture from the air.

**ICCT International Council on Clean Transportation.**

**IEA International Energy Agency.**

**Indirect Land-Use Change (ILUC).**

**In-line Blending**: Real-time automated blending of biofuel with fossil fuel.

**International Blending Mandate:** Legal requirements that mandate fuel suppliers to blend specific percentages of biofuels into gasoline or diesel.

**ISCC (International Sustainability and Carbon Certification).**

**IRR (Internal Rate of Return).**

**LCFS (Low Carbon Fuel Standard)**: A regulatory framework that incentivizes fuels with lower carbon intensities.

**LCOF (Levelized Cost of Fuel)**: The average cost per unit of energy produced, accounting for capital, operational and maintenance costs over a system's lifetime.

**Lifecycle Emissions**: Total emissions from production, processing, transportation and combustion of a fuel.

**Lignocellulosic Biomass:** A renewable resource made up of plant matter, primarily consisting of hemicellulose, cellulose and lignin.

**Lignocellulosic Feedstocks:** Renewable plant-based materials, primarily composed of cellulose, hemicellulose and lignin, that serve as a raw material for various applications like biofuel production, pulping and other biomaterial processes.

**Lignocellulosic Fuels:** Biofuels produced from lignocellulosic biomass, a renewable and abundant source of organic material.

**Lignocellulosic Materials:** Plant-based materials primarily composed of cellulose, hemicellulose and lignin.

# M–S

**Methane Slip**: Unburned methane emissions from combustion engines, especially relevant for biogas applications.

**Non-Governmental Organization (NGO).**

**NOx:** A shorthand term for a group of nitrogen oxides, primarily nitrogen dioxide ($NO_2$) and nitric oxide ($NO$).

**NPV Net Present Value.**

**OEM: Original Equipment Manufacturer.**

**Peatland Forests:** Forests that grow on peatlands, which are waterlogged ecosystems where dead organic matter accumulates over time.

**Phase Separation**: Occurs when water causes ethanol to separate from gasoline.

**Photobiological Systems:** A biological system that responds to light, encompassing a wide range of phenomena from photosynthesis to visual perception.

**Photosynthetic Fuels:** Produced through a process that mimics natural photosynthesis, using sunlight, water, and carbon dioxide to generate fuels like hydrogen, ethanol or other organic compounds.

**Pilot Project**: A small-scale test used to validate technical and economic assumptions before wider rollout.

**PM:** Refers to particulate matter, which is a complex mixture of solid particles and liquid droplets found in the air.

**Pyrolysis Oil:** A liquid fuel produced by the rapid heating of organic material in a low-oxygen environment.

**R&D Research and Development.**

**RIN (Renewable Identification Number)**: A tracking number used in the U.S. renewable fuel standard (RFS) system to certify renewable fuel production and compliance.

**Renewable Diesel:** Also known as hydrotreated vegetable oil or green diesel, is a hydrocarbon fuel produced primarily via hydrotreating, using a wider range of feedstocks including non-food biomass.

**Renewable Fuel Standards (RFS):** A regulatory framework that mandates a certain percentage of transportation fuels to be renewable.

**RSB (Roundtable on Sustainable Biomaterials).**

**Selective Catalytic Reduction (SCR):** An emission control technology that reduces nitrogen oxides (NOx) in vehicle exhaust by converting them into harmless nitrogen and water.

**SAF (Sustainable Aviation Fuel):** Renewable fuel for aircraft, derived from feedstocks like HEFA oils, alcohols, or waste gases.

**Standard Operating Procedures (SOP).**

**Splash Blending:** Manual mixing of fuels, often done at terminals or distribution points.

**Synthetic biofuels:** Fuels produced from biomass or other sources using chemical or thermal processes.

# T–Z

**Thermochemical Conversion:** A process that uses heat and chemical reactions to transform organic matter, like biomass, into useful products like biofuels and energy.

**Technology Readiness Level (TRL):** A metric used to assess the maturity of a technology, from lab prototype (TRL 1–3) to commercial deployment (TRL 9).

**Thermal Cracking:** A refining process that uses heat to break down large hydrocarbon molecules into lighter, usable fuels.

**Upstream/Downstream:** "Upstream" refers to activities related to raw material collection and processing; "downstream" refers to fuel refining, distribution and end use.

**Well-to-Wheel (WTW):** A comprehensive accounting of all emissions and energy use from resource extraction to vehicle propulsion.

# Bibliography

https://www.acciona.com.au/updates/stories/hvo

Agarwal, A. K. (2007). *Biofuels (alcohols and biodiesel) applications as fuels for internal combustion engines. Progress in Energy and Combustion Science, 33(3), 233-271.*

Bridgwater, A. V. (2012). *Review of fast pyrolysis of biomass and product upgrading. Biomass and Bioenergy, 38, 68-94.*

Chisti, Y. (2007). *Biodiesel from microalgae. Biotechnology Advances, 25(3), 294-306.*

Demirbas, A. (2009). *Political, economic and environmental impacts of biofuels: A review. Applied Energy, 86, S108-S117.*

Fargione, J., Hill, J., Tilman, D., Polasky, S., & Hawthorne, P. (2008). *Land clearing and the biofuel carbon debt. Science, 319(5867), 1235-1238.*

Hill, J., Nelson, E., Tilman, D., Polasky, S., & Tiffany, D. (2006). *Environmental, economic, and energetic costs and benefits of biodiesel and ethanol biofuels. Proceedings of the National Academy of Sciences, 103(30), 11206-11210.*

Naik, S. N., Goud, V. V., Rout, P. K., & Dalai, A. K. (2010). *Production of first and second generation biofuels: a comprehensive review. Renewable and Sustainable Energy Reviews, 14(2), 578-597.*

Ragauskas, A. J., Williams, C. K., Davison, B. H., Britovsek, G., Cairney, J., Eckert, C. A., ... & Liotta, C. L. (2006). *The path forward for biofuels and biomaterials. Science, 311(5760), 484-489.*

Sanders, J., Scott, E., Weusthuis, R., & Mooibroek, H. (2007). *Bio-refinery is the bio-inspired process of bulk chemicals. Macromolecular Bioscience, 7(2), 105-117.*

Searchinger, T., Heimlich, R., Houghton, R. A., Dong, F., Elobeid, A., Fabiosa, J., ... & Yu, T. H. (2008). *Use of US croplands for biofuels increases greenhouse gases through emissions from land-use change. Science, 319(5867), 1238-1240.*

Surmacz, T., Surmacz, T., Wierzbiński, B., Kuźniar, W., & Witek, L. (2024). *Towards Sustainable Consumption: Generation Z's Views on Ownership and Access in the Sharing Economy. Energies, 17(14), 3377.*

What Is Agile Auditing? *https://riskpublishing.com/what-is-agile-auditing/*

Biofuel Technology Evaluation Matrix
*https://www.apec.org/publications/2010/12/biofuel-costs-technologies-and-economics-in-apec-economies*
*https://www.apec.org/docs/default-source/publications/2010/12/biofuel-costs-technologies-and-economics-in-apec-economies/210_ewg_biofuel-production-cost.pdf?sfvrsn=72803e82_1*
*https://www.sciencedirect.com/science/article/pii/S0306261921014690*

Cost Breakdown of Biofuel Implementation
*https://www.apec.org/docs/default-source/publications/2010/12/biofuel-costs-technologies-and-economics-in-apec-economies/210_ewg_biofuel-production-cost.pdf*
*https://www.iea.org/reports/technology-roadmap-biofuels-for-transport*
*https://www.irena.org/publications/2016/Oct/Innovation-Outlook-Advanced-Liquid-Biofuels*

Risk Assessment Heatmap
*ISO 31000:2018 Risk Management—Guidelines, informed by bioenergy-specific analyses from the U.S. Forest Service (Risk Management Consideration in the Bioeconomy), IEA Bioenergy Task 40 (Deployment of Biofuels – Risks and Risk Mitigation), and the UNEP Bioenergy Decision Support Tool. Risk categories and scoring reflect common challenges observed in industrial biofuel implementation, including supply chain reliability, equipment compatibility, and regulatory variability.*
*https://www.fs.usda.gov/research/treesearch/55507* *https://unepccc.org/*

Schematic Biofuel Storage System
*https://afdc.energy.gov/files/u/publication/biodiesel_handling_use_guide.pdf*
*https://19january2017snapshot.epa.gov/ust/alternative-fuels-and-underground-storage-tanks-usts_.html*
*https://www.researchgate.net/figure/Biodiesel-storage-tank-for-corrosion-analysis_fig1_351027554*

Phased Biofuel Implementation Timeline
*https://www.ieabioenergy.com/*
*https://www.irena.org/publications/2016/Oct/Innovation-Outlook-Advanced-Liquid-Biofuels https://unepdtu.org*

Ethical Risk Zones in Biofuel Supply Chains: *FAO (Food and Agriculture Organization) reports on land use change and rural displacement in biofuel projects (e.g., FAO 2021: Biofuels and Land Rights), OECD/IEA publications on food vs. fuel tradeoffs and water stress in energy crop cultivation, Case studies from the World Bank, CIFOR, and academic literature on soy, palm oil, and cassava biofuel initiatives, Principles from energy justice frameworks in energy transition literature (e.g., Sovacool et al., 2017)*

Palm Oil in Southeast Asia
*https://openknowledge.fao.org/server/api/core/bitstreams/e51e0cf0-4ece-428c-8227-ff6c51b06b16/content*

Case Example: Soy Biodiesel Expansion in Brazil
*https://news.mongabay.com/2007/08/biofuels-driving-destruction-of-brazilian-cerrado/*

Community-Scale Bioethanol in Kenya:
*https://www.fao.org/in-action/global-bioenergy-partnership/news-and-events/events/events-detail/bioethanol-for-clean-cooking--opportunities-for-improving-the-sustainability-of-local-value-chains/en*

SAF Blending at Schiphol Airport (Netherlands):
*https://www.neste.com/news/neste-supplying-sustainable-aviation-fuel-to-emirates-for-flights-from-amsterdam-airport-schiphol*

Example: Financial Scenarios – 50 MGPY Renewable Diesel Plant
*https://ww2.arb.ca.gov/resources/documents/lcfs-data-dashboard*
*https://www.ers.usda.gov/publications?topicId=14835*

RED II
*https://joint-research-centre.ec.europa.eu/welcome-jec-website/reference-regulatory-framework/renewable-energy-recast-2030-red-ii_en*

RenovaBio
*https://www.sugarcane.org/sustainability-the-brazilian-experience/renovabio/*

Horizon Europe
*https://research-and-innovation.ec.europa.eu/funding/funding-opportunities/funding-programmes-and-open-calls/horizon-europe_en*

Innovation Fund
*https://climate.ec.europa.eu/eu-action/eu-funding-climate-action/innovation-fund_en*

Future Made in Australia Innovation Fund
*https://arena.gov.au/funding/future-made-in-australia-innovation-fund/*

Fulcrum BioEnergy's Sierra BioFuels Plant:
*https://www.iea.org/energy-system/transport/aviation*
*https://fulcrumnorthpoint.com/*
*https://www.energy.gov/lpo/loan-programs-office*

Levelized Cost of Fuel (LCOF) by Technology (2024 estimates)
*https://greet.anl.gov/*
*https://www.iea.org/reports/renewables-2023/transport-biofuels*
*https://www.energy.gov/eere/bioenergy/articles/2023-multi-year-program-plan*

U.S. Inflation Reduction Act
*https://home.treasury.gov/news/press-releases/jy1830*

India's Ethanol Blending Program
*https://mopng.gov.in/en/refining/ethanol-blended-petrol*

Carbon Pricing Schemes (e.g., California LCFS)
*https://ww2.arb.ca.gov/our-work/programs/low-carbon-fuel-standard*

Case Study: A U.S. cellulosic ethanol project:
*U.S. Department of Energy (DOE) Bioenergy Technologies Office (BETO) Multi-Year Program Plan (2023), California Air Resources Board (CARB) LCFS Program Updates, IEA Bioenergy Task 39 reports on advanced biofuels and market uptake, Industry interviews and reporting in outlets like Biofuels Digest, Renewable Energy World, and Greentech Media,*

Sustainable Aviation Fuel (SAF) Integration in Regional Aviation
*https://avitrader.com/2023/12/05/nac-collaborates-with-atr-and-totalenergies/*
*https://www.nac.dk/nac-collaborates-with-atr-and-totalenergies-to-offer-the-equivalent-of-6-saf-sustainable-aviation-fuel-on-nacs-atr-delivery-flights/*

Biofuel Risk Matrix:
*https://www.iso.org/standard/65694.html*
*https://www.energy.gov/eere/bioenergy/bioenergy-technologies-office*
*https://www2.nrel.gov/research/publications*

Feedstock Quality Variation vs. Equipment Downtime Chart:
*https://www.mdpi.com/1996-1073/9/3/203*
*https://biomassmagazine.com/articles/feedstock-variability-causes-conseque
nces-and-mitigation-of-biological-degradation-19639*
*https://bioenergy.inl.gov/Journal%20Articles/Understanding%20biomass%
20feedstock%20variability.pdf*

Multi-Layered Risk Management Framework:
*https://www.icheme.org/media/9263/xxii-paper-49.pdf*
*https://www.jiem.org/index.php/jiem/article/view/2812*
*https://www.bakerrisk.com/services/biofuel-risk-management-services*

LanzaTech, with the Support of Danone, Discovers Method to Produce
Sustainable PET Bottles from Captured Carbon, 27 May 2022,
*https://advancedbiofuelsusa.info/lanzatech-with-the-support-of-danone-disc
overs-method-to-produce-sustainable-pet-bottles-from-captured-carbo*n

Anellotech's Bio-TCat™ technology is ready for commercialization, 15.
12.2021,
*https://www.ifpenergiesnouvelles.com/article/anellotechs-bio-tcattm-technol
ogy-ready-commercializatio*

LanzaTech Microbial Conversion of $CO_2$ to Ethanol
*https://lanzatech.com/meet-the-microbes-powering-a-circular-carbon-econo
my/*

Clariant Sunliquid® Plant (Romania),
*https://www.clariant.com/en/Corporate/News/2022/06/Clariant-pro-
duces-first-commercial-sunliquid-cellulosic-ethanol-at-new-plant-in-Po-
dari-Romania*
*https://www.clariant.com/en/Corporate/News/2018/09/Groundbreak-
ing-for-Clariantrsquos-sunliquidreg-cellulosic-ethanol-plant-in-Romani-
anbsp*
*https://www.clariant.com/en/Corporate/News/2021/10/Clariant-completes
-construction-of-first-commercial-sunliquid-cellulosic-ethanol-plant-in-Pod
ari-Rom*

Technology Integration Challenges and Enablers: *DOE BETO Multi-Year
Program Plans (2023), IEA Bioenergy Task 42, Commercial case examples
(Clariant, LanzaTech, and industrial AI applications), Industry best prac-
tices in AI deployment, synthetic biology scale-up, and catalytic process design*

Integrated Risk Management Zones: *Adapted and synthesized by the au-
thor from IEA (2023), U.S. DOE BETO (2023), and public case documentation.*